QUALITY SYSTEMS

ASBC HANDBOOK SERIES

Practical Guides for Beer Quality

QUALITY SYSTEMS

Charles W. Bamforth
Distinguished Professor Emeritus
University of California, Davis

ASBC
AMERICAN SOCIETY OF BREWING CHEMISTS

Front cover: Courtesy atdr, Shutterstock.com—Reproduced by permission.

Library of Congress Control Number: 2019940646
International Standard Book Numbers:
Print: 978-1-881696-34-6
Mobi eBook: 978-1-881696-35-3
ePub: 978-1-881696-42-1

Second printing 2021

Printed in the United States of America on acid-free paper

American Society of Brewing Chemists
3340 Pilot Knob Road
St. Paul, Minnesota 55121, U.S.A.

For my mother, Edith (born in 1920 and still in charge!),
and my wife, Diane.
Without either of them, I would have achieved very little.

This Book and This Series

This is the fifth volume in a six-part series addressing quality issues of beer. There are, of course, a number of books and numerous articles that address such matters. This particular series might be viewed as a set of user manuals, much like the handbook that accompanies your car.

In each book I seek to address individual quality issues from a standpoint of

- the basic underpinning science (without getting too complex and grinding the grist too fine)
- the practicalities of the issue pertaining to brewing and its associated activities
- quality assurance and quality control parameters
- a troubleshooting guide (or key control point summary)

Acknowledgment

I am very grateful to Steve McKinley for his excellent artwork.

Suggestion

For those readers who do not have a fundamental understanding of the science and technology of brewing, I suggest that a basic text on beer brewing be read. There are some suggestions under "Further Reading."

Contents

1. So What Is Quality? 1
2. Systematic Quality 7
3. Some Basic Statistics 13
4. Standard Methodology 19
5. Assuring Quality 29
6. Critical Stuff 37
7. Setting Up a Quality Program 47
8. A Summary of the Key Points 67
Further Reading 69
Index 73

QUALITY SYSTEMS

So What Is Quality?

Chapter 1

Give them quality. That's the best kind of advertising.
—Milton S. Hershey

My long-suffering wife, Diane, has a pretty good idea what I am likely to cook for myself if she happens not to be home. It will be Charlie's chili, which bears no resemblance to that of anybody else on the planet. Having warmed up a bit of oil in a pan, I will brown my ground beef and then flavor it with salt, black pepper, chili powder, and lashings of Worcestershire sauce. Then I add a very large portion of onions, which must be sliced into rings. Having mixed things up, it is now a matter of adding a can of Heinz Baked Beans (no other brand will do) and stirring and simmering for 45 minutes. I then devour accompanied by crusty bread.

A good many of you will not at all like the sound of this. That's okay. All that matters is that it is a dish that delights me. Minced turkey does not work. Onions in tiny pieces do not work. Other brands of beans do not work. I know what I like. I know what I want—every time, no surprises.

This is my definition of quality: A product that meets my expectations every time. No deviations. It can be put in many ways:

- the extent of agreement between expectation and realization
- the match between what you want and what you get
- the achievement of consistency and the elimination of unwanted surprises

And my favorite

- the supply of goods that do not come back to customers who do

Just as it applies to my favorite home-cooked concoction, so does it apply to beer and pretty much anything else that exists in our world. Oh sure, there is room for some freedom of expression in certain activities, matters artistic, for example. However, there is only one Mona Lisa. Moreover, if one buys a print from that painting, then we do not actually want it sometimes to have a mustache over those tender lips and beneath that enigmatic gaze.

So let us hop back to my culinary concoction. Yes, there is a very real probability that you would hate it. All the more for me. Likewise, I could probably list any number of things that you like that turn my stomach: rare steak, avocado, peanut butter and jelly sandwiches, medical dramas on television. Good luck to you if these are what get you high. I am sure you know what you expect from each of them and whether they meet your expectations. Just do not expect me to want to join you.

The definition of quality, of course, extends to beer. I am fairly catholic in my tastes when it comes to beer and will at the very least take a sip of pretty much anything (except beer containing animal reproductive glands or the secretions thereof). There are some products I adore (e.g., good examples of black IPA, maibock, English cask bitter, hefeweissen, Irish stout in a rustic Dublin bar). There are others I will quaff and be grateful for when the occasion demands it (e.g., a North American lager on a sweltering California day). Some beers I am not a fan of (e.g., Rauchbier, sours, ridiculously turbid stuff). Some I find distasteful (notably nitrokegs, other than when it is an Irish stout). However, it really matters only to me. That is my taste. It is up to us all to decide what we do and do not savor. The simple requirement is that the drink should meet our expectations. I expect a hefeweissen to be rich in clove and banana and absent a slice of lemon. You might favor a brand labeled as that style but that does not have those traditional characteristics. No problem—as long as it satisfies your wishes every time.

The essential thing is that you should not seek to embarrass, denigrate, or belittle somebody whose favorite beer is the antithesis of what you prefer.

In the four previous books in this series, we encountered various examples of how to achieve consistent quality. In this volume, we take a more overarching view of the issue.

Quality Is Driven from the Top

Robin Manners (Fig. 1-1) was a very gracious man. He lived up to his surname. I remember once, as the young research manager for Bass Brewers, having been asked to chauffeur him to his Staffordshire home following a dinner in Blackpool. For 3 h he bombarded me with questions about our beers and our breweries, teasing out my opinions and thoughts. This was more than a decade before he became the High Sherriff of Staffordshire in 1998. Robin Manners cared. His words are imprinted on my brain: "There are two things that are of the top priority in our company, Charlie. The first is our people, the second is quality. And if we look after the former, they will look after the latter."[1]

Fig. 1-1. Robin Manners. (Courtesy Judith Manners, with thanks to Harry White—Reproduced by permission)

[1] In 2013, I exchanged emails with Robin Manners, who died too young in 2015. I reminded him of what he had said to me about people and quality. He replied, "With the benefit of hindsight and experience, I would also have added make sure that your people are doing a job that they enjoy, that they have been properly trained to do it and that they understand why it is important to the Company. If they have potential to go higher in the organization make sure that they have, in advance, the development training they need to do the next job up."

At a stroke, we focus on two of the great truths of any quality organization: Quality is driven from the top, and it depends on the efforts of *all* of the people in that company.

Beware bean counters. I recall a former student of mine calling me in distress after a visit to his brewery from a legendary brewing magnate. It seems that the attack had been relentless, as everyone from the brewmaster on down was on the receiving end of an onslaught of criticism, a blow-by-blow account of what they were doing wrong. I soothed "my man" and pointed out that this really was a very high-class problem. "If the guy whose name is on the label is so passionate about excellence, that is something to be excited about. Sure, he could probably approach things in a kindlier way, but it is far better for him to come in all guns blazing than to have a bean counter thousands of miles away, scrutinizing numbers in isolation, someone who worries primarily about the bottom line and who might even refer to beer as 'liquid.'"

Think of any "quality forward" organization, and you will find a special person at the helm. In brewing, I would mention the likes of Ken Grossman of Sierra Nevada (Fig. 1-2) and Tim Cooper of Coopers

Fig. 1-2. Ken Grossman. (Courtesy Ken Grossman, with thanks to Melissa Cafferata—Reproduced by permission)

Fig. 1-3. Tim Cooper. (Courtesy Tim Cooper—Reproduced by permission)

Brewery (Fig. 1-3). Take a pilgrimage to the Sierra Nevada breweries in Chico, California, and Mills River, North Carolina, or the Coopers Brewery in Adelaide and prepare to be wowed.

Grossman and Cooper believe in people. They know that quality is not the province of some white-coated, scientifically minded boffins in an ivory tower wielding pipettes, swabs, and spectrophotometers. Analysis is important, as we will emphasize. However, it is attitude that is paramount. The right outlook from everyone. Quality is everybody's responsibility. It is not something safeguarded by a separate department regarded as some form of gestapo by the rest of the workforce with a perception of being there to hinder and hand out blame and criticism.

Quality is a state of mind. Quality should be at the heart of and in the mindset of every employee of a company.

Systematic Quality

Chapter 2

The phone rang. On the other end was the managing director of a brewery in the Bay Area, inquiring whether I did quality audits. I answered in the affirmative. Did they want an expensive one or a cheap one? They replied the latter and wondered what it would comprise. I said that I would do 3 half days. On the first morning, I would inspect their records and their standard operating procedures. On the second, I would tour the facility and make observations. On the third morning, I would report my findings. The deal was struck.

I duly turned up on the first day and was presented with two boxes, a large one and a small one. In the large one was an abundance of loose-leaf pages, completely randomized with a plethora of numbers on them that meant very little. In the small box was a solitary operating procedure, which read, "To tighten mill, turn knob towards Golden Gate Bridge." The second day confirmed my worst impression: The brewery was a mess—structurally, functionally, and hygienically. The meeting on the third day was uncomfortable. The brewery is long gone.

If you are going to achieve consistency, you must do things consistently. In other words, you must specify things: the materials, the process, and the products. I speak of the "holy book": the quality manual. Likely these days it is all online. Whether it is hard copy or electronically encompassed, it needs to exist, be kept up to date, and be slavishly followed by all those who truly believe in quality. In a nutshell, you must

- define the products in sufficient detail to know what you are seeking to produce and to ensure that you can do it time in, time out;
- define every single material used in that brewery; above all, anything that will end up in the beer but, importantly, anything that forms a key part of the process, from process aids to cleaning agents; and
- define exactly how all the processes are to be carried out, not only the actual brewing, fermentation, and downstream processes, but also the "surrounding" processes, such as cleaning and sterilization. These are the standard operating procedures (SOPs) (Box 2-1).

The test for whether you have the necessary detail is the double-decker bus analogy. If you are the brewer at the Dog's Posterior Brewing Company, celebrated for your wonderfully consistent and acclaimed Bottom Line Ale and Bottom Dollar Lager, and you have the unfortunate experience of being run over by the aforementioned bus, will somebody else be able to gen up on the specs and the SOPs and, in a respectable and respectful period of time, carry on the show?

Good Manufacturing Practice

The phrase "good manufacturing practice" (GMP) represents the basis of any production process or entity dedicated to delivery of quality products. The United States Food and Drug Administration demands that all breweries delivering a product for sale have a functioning GMP program.

In considering the scope of GMP, the brewer needs to embrace the facility and the grounds in which it lies, the equipment and materials employed, the approaches to matters sanitary, the processes and their control, and, finally (though by no means last), the people.

In relation to this, the Food Safety Modernization Act (FSMA) insists upon properly documented SOPs and records from incoming raw materials onward. The reader is referred to the websites of the Brewers Association (www.brewersassociation.org) and the Master Brewers Association of the Americas (www.mbaa.com) for links to more guidance on the extent of the regulations.

Box 2-1. Example of a Standard Operating Procedure (SOP)

Process: Roller Mill Operation
Location: RMI Brewery, Winery, and Food Pilot Facility 1109a
PPE Required: Safety goggles, respirator
Created: 15 Nov 2018 KAS
Last Revision Date: 19 Nov 2018 JAW

Preparation

1. Weigh malt into bags or buckets to a weight that you can safely and comfortably lift to chest height.
2. Remove the grain hopper cover and check for cleanliness.
3. Tap the mill discharge opening to remove any dust.
4. Ensure the mill hopper slide is closed.
   a. Slide lever should be completely vertical and pointed at "0".
5. Secure a bag or bucket sock to the mill outlet depending on amount to be milled.
   a. Less than 10 kg use buckets.
   b. More than 10 kg mill into empty grain bags.

Note: Milled grain takes up more space than un-milled. Plan accordingly and don't overfill.

6. If using a bucket, insert a bucket extender and a bucket dolly. Center the bucket under the mill outlet and place the grain discharge sock into the bucket.

Mill

1. Slowly pour grain into the hopper.
2. Brush any grain off the screen into the hopper with a brush, looking for foreign material.
3. Turn on the mill by pressing the black "I" button on the front of the mill.
4. Slowly open the hopper slide to "15".
5. If the amount loaded into the hopper will fit into the receiving vessel, mill all the malt in the hopper and allow the mill to run for one minute after all grain has been milled.
6. If the hopper is loaded with more malt than will fit into one receiving vessel, monitor the level in the receiving vessel and do not overfill.
   a. To stop milling, close the hopper slide before turning off the rollers.
   b. Do not allow the rollers to stop while there is grain in the mill; the rollers will bind on restart.
   c. Malt can remain in the hopper as long as the hopper slide is closed.
   d. Swap receiving vessel and restart milling operation.
7. Once all grain has run through, close the hopper slide to "0" and turn off the mill.
8. Tap the mill discharge opening to remove any dust.
9. Remove the bag or bucket from the mill.
10. Clean all dust from under the mill and near where grain was weighed. Do not leave a mess.

Extracted from the SOPs for the pilot brewery at the University of California, Davis. Courtesy Joe Williams—Reproduced by permission.

GMP and FSMA are particularly focused on health and embrace tightly the approaches of hazard analysis and critical control points (HACCP), which we will tackle in some detail in Chapter 6.

ISO

There are those across the globe who formalize their general approach to achieving a quality product in something called ISO 9000 and ISO 9001. ISO stands for International Organization for Standardization (www.iso.org). In short, the program sets out the range of requirements that must be met for a company to use the ISO logo as a badge of honor on its paperwork to indicate that its employees strive to "meet the needs of customers and other stakeholders while meeting statutory and regulatory requirements related to a product or service."

ISO specifies the quality system that proves the capability to manufacture and supply product to an established specification. It defines the activities that must be controlled but not how they are to be controlled.

The following are key elements of ISO 9001. Even for those companies not seeking accreditation, these points represent a very useful map to addressing quality in a structured and systematic manner. Let us walk through the items on this "shopping list" one by one.

A quality policy, usually signed by the senior executive, which speaks of the company's commitment to quality and to meeting the needs of customers. Figure 2-1 illustrates a good example of this. We are in the territory of what Robin Manners told me (see Chapter 1).

A quality manual that outlines and maps the scope and elements of the quality management system and how these may be cross-referenced to relevant procedures and work instructions. Written procedures for all key operations that affect product quality are mandatory.

Systems for ensuring uniform use of updated documentation, including specifications, control diagrams and schemes, codes of practice, operating manuals, and HACCP. Everyone needs to sing to the same hymn sheet. Things change. You might decide that the bitterness of a given brand needs to come down or go up. You might change from a silica gel stabilizer to polyvinylpyrrolidone (PVPP) (see *Color and Clarity*, the fourth volume in this series). You might have switched out your lauter tun for a mash filter. The documentation needs to change, and everyone who has a copy of the holy book needs to update the relevant pages.

Secure records to demonstrate that the system is working effectively and meets legal and contractual requirements. More docu-

SAN MIGUEL BREWERY INC.

QUALITY POLICY

San Miguel Brewery Inc. is dedicated to delight its customers with products and services of superior quality.

True to our commitment to Quality, SMB shall adhere to the following action principles:

- Consistently manufacture products in the tradition of San Miguel's brewing excellence and ensure its availability to our customers
- Optimize our capabilities based on performance and benchmarks while strengthening alignment and deployment of strategies for effective execution
- Enhance the customer management system to meet and anticipate customer needs as well as secure customer relationships
- Continuously improve our quality management system through timely and effective implementation of measures and introduction of appropriate innovation
- Observe strict compliance to statutory, regulatory and other applicable requirements
- Nurture a work environment that drives employees toward stronger commitment to quality and high performance
- Promote stronger collaboration with business partners to improve our collective performance

Quality is a way of life in SMB. It is everybody's concern. Together, we shall continue to uphold San Miguel's reputation as the "Home of Quality Products and People."

ROBERTO N. HUANG
President

May 30, 2013

Fig. 2-1. Example of a quality policy. (Courtesy San Miguel Brewery—Reproduced by permission)

mentation, I am afraid, and not necessary if you are not seeking the ISO standard. Sure, you need to keep within the law and adhere to contracts, but records demonstrating this need not be formalized.

A schedule of meetings to review the quality system. Meetings are inevitable in any organization involving more than a single person! I worked for someone once who said, "We should have a committee to discuss this and that committee should comprise just me." Yes, it is good for all relevant parties to discuss how well they are doing, but the smaller the company the more likely they are to be about quality per se and not the quality system.

Training of all staff whose jobs impact product quality. I believe that everyone's job affects quality. If the janitor is not ensuring that the lavatories are functioning, things will grind to a halt quickly. Again, the important thing is that people are properly trained to do their specific jobs.

Control of purchasing of all materials and services that can affect product quality, e.g., lists of approved suppliers and auditing regimes. Irrespective of the size of a company, a good rapport with suppliers is needed. A golden rule is to have at least two suppliers for any material, in case one of the suppliers dries up for whatever reason.

Calibration programs for equipment according to recognized standards. Any instrumentation that a brewer uses to measure any aspect of the raw materials, process, and product needs to be reliable. We will come to this in Chapter 4. Moreover, we can include in that category the human senses of smell and taste.

Internal audits to ensure that the system is effectively implemented and maintained. For a small company, this might mean bringing in an expert outsider to run the rule over the system (see my experience in the Bay Area mentioned earlier).

Systems for handling nonconforming product. Even with the best intentions in the world, things will go wrong and a product will not be fit for the market. How are you going to prevent it getting out there?

Systems for handling complaints. Who is going to respond to the letters of complaint? To what extent do you trust the complainant? When I was a quality assurance manager with Bass, it was amazing that someone who complained about finding a bit of glass in one of our beers (often a can!) was desperately unlucky enough to simultaneously discover glass in the beers of several other companies. It is a great idea to be a part of a consortium of brewers who can exchange details of serial complainers and much more besides.

Systems for adjusting standard procedures in order that problem recurrence is avoided. Complaints can be genuine. Maybe there was a foreign body in a can. How is it likely to have got in there and what can you do in the process to prevent it from occurring again? Make sure you change the SOPs in relation to anything you change.

Some Basic Statistics

Chapter 3

It is essential that any brewing operation is founded on reliable data. We must be able to measure things if we are going to interpret what is going on, respond to problems, and make whatever corrections are necessary in our single-minded determination to assure consistent excellence in our beers.

We need two things if we are to make reliable measurements: carefully defined methods and the ability to understand the statistical significance of what we are measuring. In this chapter, we address statistics.

Mean

Let us start with the mean. You might call it the average. Let us say that we found ourselves in a small island nation that boasted 15 breweries, each of which produced four types of beer. Armed with our trusty gas chromatograph (what self-respecting brewing scientist does not travel with one of those in his or her luggage?), we measure the alcohol content by volume (ABV) in each of those beers and we find these values:

5.0 (9 beers), 3.9 (3 beers), 7.2 (2 beers), 2.5 (1 beer) , 5.1 (9 beers), 5.5 (13 beers), 6.2 (5 beers), 9.0 (1 beer), 1.3 (1 beer), 8.2 (1 beer), 5.2 (8 beers), 7.8 (1 beer), 4.3 (6 beers).

We get the mean by adding them all together and dividing the sum by the total number of measurements (60), giving us a value of 5.26.

If we call the mean *m,* then we can explain this calculation using a simple formula:

$$m = \frac{1}{n} \Sigma x_i$$

where Σ is "the sum of" (i.e., all the values added together), x_i refers to each individual measurement, and n is the total number of measurements.

Range

Now, if you look at these 60 numbers, you can see that the average ABV is at the middle of quite a huge range of strengths, with a low of 1.3% (no thank you) and a high of 9.0% (okay, but just the one, please). Just by knowing the mean, we really don't know much about our population of samples. The easiest way is to plot them out (Fig. 3-1). You can then see that actually the greatest number of beers have an ABV of 5.5%, which is above the mean. However, the greater preponderance of beers with a lower ABV renders the mean a little less than this.

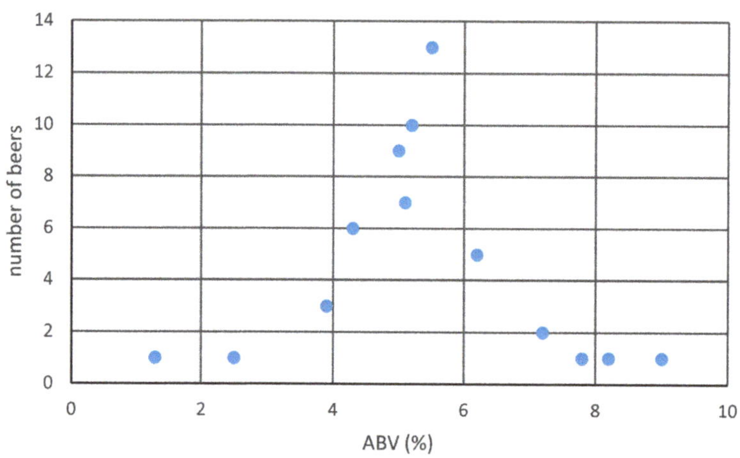

Fig. 3-1. Data distribution about a mean. ABV = alcohol content by volume. (Courtesy C. W. Bamforth—© ASBC)

So how can we express the range in a formula?

The mean absolute deviation (MAD) is given by

$$\text{MAD} = \frac{1}{n}\,\Sigma\,|x_{\text{i}} - m|$$

The | notation indicates that the + and – values are all rendered positive, irrespective of whether the value is lower or higher than the mean. In other words, what we have is the total of all the differences between each value and the mean divided by the total number of measurements. Table 3-1 gives us the necessary values.

Let us put all those values in the right hand column into our MAD formula, and we get a value for MAD of 0.70.

Table 3-1.

Spread of Data in Calculating Mean Absolute Deviation[a]

ABV[b]	Difference from the Mean	Total Beers at This ABV	$\Sigma\,|x_i - m|$ for This ABV[c]
1.3	–3.96	1	3.96
2.5	–2.76	1	2.76
3.9	–1.36	3	4.08
4.3	–0.96	6	5.76
5.0	–0.26	9	2.34
5.1	–0.16	9	1.44
5.2	–0.06	8	0.48
5.5	0.24	13	3.12
6.2	0.94	5	4.70
7.2	1.94	2	3.88
7.8	2.54	1	2.54
8.2	2.94	1	2.94
9.0	3.74	1	3.74

[a] Courtesy C. W. Bamforth—© ASBC.
[b] Alcohol content by volume.
[c] x_i = each individual measurement, and m = mean.

The other way to overcome the problem of negative values is to use squared values, and now we have the mean square deviation (MSD):

$$\text{MSD} = \frac{1}{n}\,\Sigma\,[x_i - m]^2$$

Let's do that in Table 3-2. If we calculate MSD according to our formula, we get 1.3.

Table 3-2.

Spread of Data in Calculating Mean Square Deviation[a]

ABV[b]	Difference from the Mean $(x_i - m)$	Rounded Square of This $(x_i - m)^2$	Total Beers at This ABV	$\Sigma\,(x_i - m)^2$ for This ABV[c]
1.3	−3.96	15.6816	1	15.6816
2.5	−2.76	7.6176	1	7.6176
3.9	−1.36	1.8496	3	5.5488
4.3	−0.96	0.9216	6	5.5296
5.0	−0.26	0.0676	9	0.6084
5.1	−0.16	0.0256	9	0.2304
5.2	−0.06	0.0036	8	0.0288
5.5	+0.24	0.0576	13	0.7488
6.2	+0.94	0.8836	5	4.4180
7.2	+1.94	3.7636	2	7.5272
7.8	+2.54	6.4516	1	6.4516
8.2	+2.94	8.6436	1	8.6436
9.0	+3.74	13.9876	1	13.9876

[a] Courtesy C. W. Bamforth—© ASBC.
[b] Alcohol content by volume.
[c] x_i = each individual measurement, and m = mean.

Standard Deviation

And so we arrive at the term that statisticians tend to use and that you are more likely to have encountered, namely, "standard deviation." We go to the extra stage of taking the square root of what we have just calculated. By doing this, we get a number that is of the same order of magnitude as the values that we are actually measuring, in this example ABV.

The formula for standard deviation, usually referred to as σ, is

$$\sigma = \sqrt{\frac{1}{n-1}\Sigma\,[x_i - m]^2}$$

You will note that rather than *n*, we are now using $n - 1$. Don't ask! It is no biggie. So our value for σ in this example is 1.1.

If you look at our ABV range (a low of 1.3% and a high of 9.0%), you can see that a standard deviation of 1.1 is not the same value you would get from subtracting the lowest value from the highest one. In short, a small standard deviation signifies that the mean is the average of very similar values. A large standard deviation indicates substantial data spread. The fact that the greater proportion of values in our data set were in the region close to the mean explains why the standard deviation is a smaller number than it would have been if, say, we had a range of beers with equal numbers of samples at each ABV. If we had the latter situation, then the standard deviation would be very close to being the number that, if you subtracted from or added it to the mean value, would have given you the two values at the extremes of the data range.

Normal Distribution

For many data sets, particularly if there are very large numbers of samples, the plotted values adopt what is known as the "normal distribution" (Fig. 3-2). For example, if you measured the ABV of all the beers across the world, then the mean would (I guess—I haven't done this) be around 4.5%. There would be genuine alcohol-free beers at one end and products like Snake Venom (67.5% ABV) at the other. However, the vast majority (I venture) would be in the range of 0.5 to 8.5% ABV, with a progressively decreasing minority down from and above the extremes. The height or weight of the male and female populations is a great example of a normal distribution, though the mean for each is higher than it was a century ago. The shape of the data set is probably the same, but heights on average have increased and weights most certainly have. However, let us stick to beer.

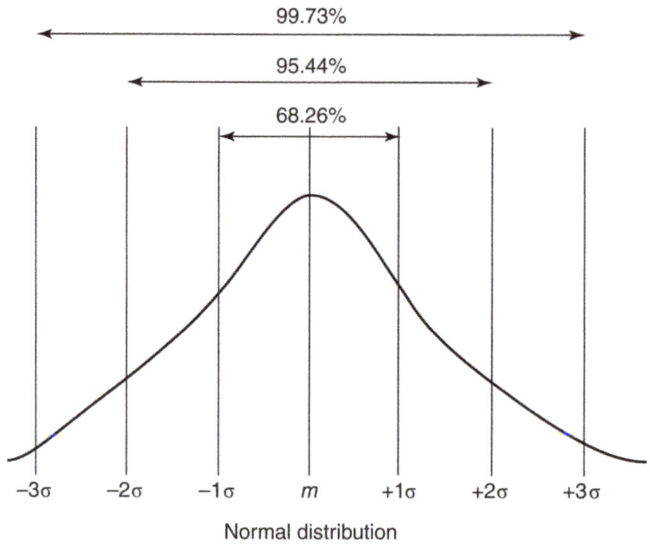

Fig. 3-2. A normal distribution. σ = standard deviation, and *m* = mean. (Courtesy C. W. Bamforth—© ASBC)

Folks call these plots "bell curves"—look at the shape. The important thing to observe is that 68% (roughly two-thirds) of the measured values are within one standard deviation (σ) above or below the mean (*m*), while 19 of 20 measurements (95%—actually just a tad higher) are within two standard deviations above or below the mean. Almost all, but not quite (99.7%), are within three standard deviations above and three standard deviations below the mean. I address the significance and application of this in Chapter 4.

Standard Methodology

Chapter 4

As a native-born Englishman, I insist upon tea as a critical component of my daily existence. Having lived in the United States for many years, I despair at the widespread inability to make a decent cup of tea. The key components are good quality tea, of course, but also *boiling* water. Having mixed the two to the desired robustness (nice and dark for me, somewhat less so for my better half), there then must be a suitable steeping period (many actually call it "mashing," others call it "brewing"), which for my preference is 3–5 minutes (Mrs. B. prefers less). Then comes another essential—if you want your tea with milk, which is how I take it, you always pour the tea into the milk and not the other way around. The proteins and tannins behave differently in the two modes, and the preferred result comes when the milk is put into the cup or mug first.

Which is quite a lengthy illustration of the key point to be addressed in this chapter: If you do the same task in different ways, you will get a different result. It applies to making tea, and equally it applies to anything you do in a malthouse or a brewery. We spoke in Chapter 2 of standard procedures. You have to do things consistently to achieve a desired result.

This applies absolutely in the methods that are used to make measurements that underpin our processes. For any analysis, we have to perform the procedure in a consistent manner. Deviate from the prescribed method and the result is very likely to be impacted.

Say a brewer is interested in controlling the level of dimethyl sulfide in the beer. To do this (as we saw in volume 2 of this series, *Flavor*), a brewer (and therefore a maltster) needs to worry about the level of the key precursor *S*-methylmethionine in the malt. It needs to be specified and measured reliably. If the maltster uses one method and the brewer another method, they will never agree. They have to use the same prescribed procedure. Moreover, this applies to everything else that needs to be measured in order to control a process and check out a raw material or a beer.

Standard Methods

Thus we have groups of methods that are the go-to resources to ensure that laboratory work is pursued consistently, no matter where a test is run.

In Germany, brewers use the methods from the Mitteleuropäische Brautechnische Analysenkommission (MEBAK, Central European Commission for Brewing Analysis; www.mebak.org). However, rather more widely employed in many parts of the world are the procedures of the European Brewery Convention (EBC; www.analytica-ebc.com), which incorporate the methods from the former Institute of Brewing. In the Americas, the go-to procedures are those of the American Society of Brewing Chemists (ASBC; methods.asbcnet.org).

In considering the EBC and the ASBC methods, the reality is that there are many similarities between them. However, there are differences, and it is important that you know what method has been used to generate a certain result. We came across one example in volume 4 of this series, namely, the measurement of color. In both sets of methods, color is estimated by measuring the absorbance at 430 nm and multiplying by a factor. However, in the EBC version, that multiplier is 25, whereas in the ASBC method, it is 12.7. Any given beer will have a higher color value if you are talking EBC units than if you are adhering to the ASBC approach. The same volume in this series highlighted that something similar occurs with haze measurement, where the multiplier is 70 times higher in the ASBC approach than in the EBC approach. In both instances, as long as you know what method is used, then you are okay. You are comparing like with like, that is, speaking a common language.

How Are These Methods Established?

First, is there a need for a new method? Let us take a hypothetical example (although perhaps I might be sowing an idea here, because this example does feature a worrisome material for which there is no ASBC method). Let us suppose that some researchers have made a detailed study of the factors that impact the levels of oxalic acid in malt. They have published a very nice paper in the *Journal of the American Society of Brewing Chemists* and highlighted how brewers and maltsters really do need to pay attention to oxalic acid. After all, we all know the problems that it can cause with beer stone, clarity, and gushing.

Being typical scientists, the authors have focused more on the introduction and the results and discussion parts of the paper and have rather hurried in putting down the precise details of their methodology. It has been written in such a loose manner that exactly what they actually did is open to interpretation. Small wonder then that folks trying to reproduce the findings have some difficulties.

Stage one then is to write the procedure in a step-by-step fashion, such that it is completely clear what needs to be done. If there are specific sources and catalog numbers of reagents, they need to be specified. The rule is that the method should be transparently clear and in a form such that anyone, anywhere, reading it and following it should be able to execute the technique exactly as intended.

The next stage is to ascertain who in the wide world is interested in the measurement of oxalic acid. These people are invited to sign up, and they will then be sent the method, together with samples to measure. Everyone will get the same samples. Everyone follows the procedure with all the samples and sends in their data to the coordinator of the trial, and the latter examines the results statistically to see just how closely everybody comes to agreement on the values for oxalic acid across the range of samples. If there is good agreement from the folks involved in the trial, then the method is published in the *Methods of Analysis*. So how do we define "good agreement"?

Accuracy and Precision

It is easiest to think of this in terms of archery (Fig. 4-1). Imagine that we have an archer who is shooting red arrows at the board and is getting them pretty close to the bull's-eye. They are arrayed around it. This archer is quite accurate. The person firing the green arrows for some reason is hitting a spot some distance from the bull's-eye, but, my word, he or she can hit the same spot every time. This person is precise. Ideally, we would like both accuracy and precision, in other words, all the arrows hitting the bull. However, if we think of this analogy in terms of what we want in malting and brewing, we would go for precision over accuracy. In the context of method development that we have been talking about in this chapter so far and saying that each arrow represents a different person carrying out a test, then if all the results are closely clustered together (precise), this is great. All we need to do is define this as the spot we are aiming for.

Let me give you a couple of examples, one trivial and one "real world."

Let us say that I am color blind (I am not, but stick with me). Let us also say that my wife, the long-suffering Diane, takes day-to-day responsibility for picking out my clothes. (She doesn't, but many people think that it would not be a bad idea.) Let us say that one day I say, "I would like to wear the red shirt," my color blindness telling me that said garment is red but it is, in fact, green. My wife, knowing of this misidentification, would mutter but successfully pick out the correct shirt. We would have a precise result. We shifted the target from the "true" and accurate situation (the bull's-eye) to the place where my expectations were.

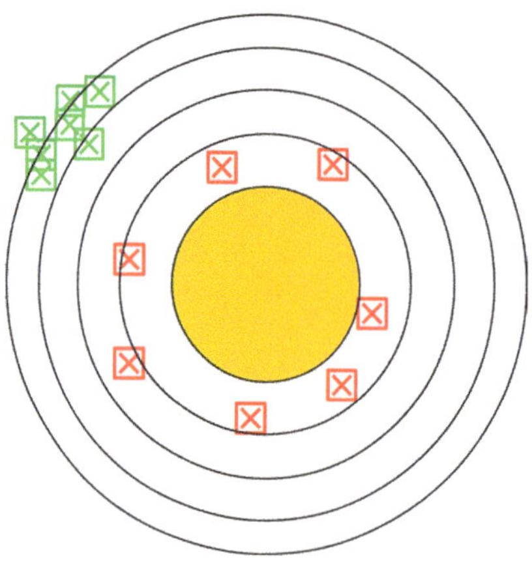

Fig. 4-1. Accuracy and precision. (Courtesy C. W. Bamforth—© ASBC)

The other example has to do with the measurement of protein in barley. For the longest time, this was done with the method of Johan Kjeldahl (Fig. 4-2), in which proteins are chemically broken down into ammonia, which is quantified. However, there was a switch several years ago to measuring protein by the method of Jean-Baptiste Dumas (Fig. 4-3), in which the protein is converted all the way into nitrogen. This method gives higher values because of the more complete combustion. Therefore, when folks switched over, the uninformed said, "The protein levels in the barley are higher." The reality was that they were not at

Fig. 4-2. Johan Kjeldahl. (Courtesy The Royal Danish Library—Reproduced by permission)

Fig. 4-3. Jean-Baptiste Dumas. (Courtesy Lumen Osad—Reproduced according to terms of CC BY-SA 3.0 license (https://creativecommons.org/licenses/by-sa/3.0/deed.en))

all higher; it was just that the method used had changed, resulting in a higher value. The solution was to readjust the specification upward a bit. In other words, the target was shifted from the existing bull's-eye to one a little ways away.

In summary, we should not be too hung up about an ultimate value, perhaps one linked back to some sacrosanct standard locked away in a vault in deepest Paris. Rather, we should know our method, know our samples, and set our specifications in relation to that method.

To bring us back to setting up the standard method, then, we are looking most closely at what the precision is. Sure, we can set up reference standards with pure solutions of oxalic acid, but it may well be that by using the extraction conditions laid down in the procedure there may not be a complete pulling out of the oxalate. Therefore, in this example, perhaps we would set our specifications based not on the solutions of oxalic acid from a bottle, but rather using oxalic acid added to our grain sample.

In any method, there are going to be sources of error. They may arise from different batches of chemicals (hence the need to be as specific as possible), unconscious deviations from the laid-down method, contaminating species in glassware or water, atmospheric conditions, or human imperfections. The list goes on. What we are seeking are methods that are sufficiently robust to overcome these matters to the greatest possible extent.

The Youden Plot

A useful tool to evaluate procedures—but also people's ability to routinely carry out existing procedures—is the Youden plot, named for William Youden. A simplified illustration is shown in Figure 4-4.

Let us think about it in the context of our oxalic acid example. Eighteen different laboratories registered interest and analyzed the two samples sent out, malt A and malt B. When the 18 pairs of results were received at headquarters, the number crunchers worked out the mean values (see Chapter 3), indicated in the figure by the values you can see for A and B with an overbar above each. Having done that, the statistician produced the circle. This represents the 95% confidence layer. Where does that come from? Think back to Chapter 3 and what I said about a normal distribution. Assuming that the measurement of

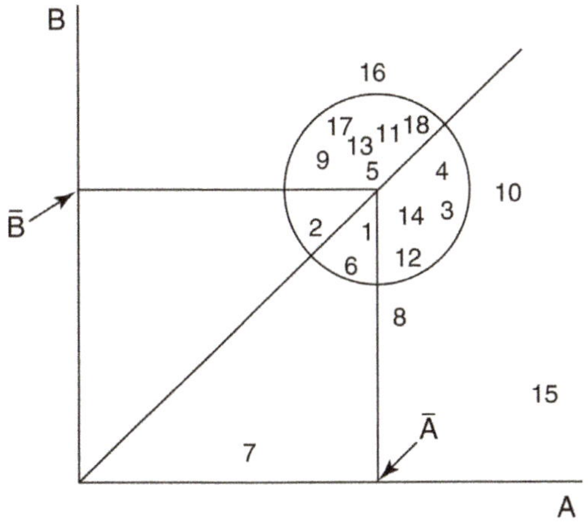

Fig. 4-4. A Youden plot. (Courtesy C. W. Bamforth—© ASBC)

oxalic acid is like other measurements and adopts a normal distribution, then we would expect 95% of the values to come within two standard deviations above and two standard deviations below the mean. In other words, that is where we would expect the bulk of the data to fall. Those laboratories whose data for these samples fall within that circle are essentially generating the same number within the range of statistical reliability. What about the others? Laboratory 16 got an outlier for sample B (they are measuring it too high), but they are in the right ballpark for sample A. Laboratory 8 is also okay on sample A but low on B. Laboratory 10 is high on A but okay on B. In other words, laboratories 8, 10, and 16 have a consistency problem. There is an even bigger problem with laboratories 7 and 15. The former is getting values that are too low for both samples. Data from laboratory 15 are all over the place, registering too high for A and too low for B.

On balance, we would say that this is not looking like too bad a method, but the preferred situation would be to have all 18 labs within that circle.

The Youden plot is just as valuable for testing the performance of people following existing methodology. Let us change the example and say that we are measuring isoamyl acetate in beer. Using the same plot as an example, we can identify which labs are having either a sporadic or a systematic problem with their measurements. Large brewing companies can use this approach to check how all the labs in all their breweries are shaping up. Alternatively, this approach can be taken with labs from separate companies measuring standardized samples sent out from a central location to see if they are reliably measuring parameters when judged against other labs. This is the basis of the ASBC Laboratory Proficiency Program (www.asbcnet.org/lab/lpp).

r and *R*

Finally, let us consider "the two r's," repeatability (*r*) and reproducibility (*R*), little *r* and Big *R*, which quantify precision.

The definition of *r* is the difference between two single results found on identical test material by one operator using the same apparatus within the shortest feasible time interval will exceed *r* on average not more than once in 20 cases in the normal and correct operation of the method.

What does that mean? It means that if you have one person making a measurement on the same sample in the same lab using the same kit repeatedly in the shortest period of time, then 19 times out of 20 the result obtained will be within this distance from the mean (19 out of 20? 95%!), that is, plus or minus two standard deviations either side of the mean. So the lower the *r* value, the more precise the analysis is.

The definition of *R* is the same but refers to the difference in results found *between* laboratories. As we have stressed in this chapter, there is going to be more inconsistency between labs (see the Youden plot example). So *R* is always bigger than *r* (Big is always more than little!).

What we are saying is that if two different labs find a result for a given sample that is within the *R* value that has been established for that method, then they are agreeing on the result.

Perhaps an example will help. Say we have the Burly Brothers malthouse furnishing pilsner malt to the Supreme Sisters brewery. The dimethyl sulfide (DMS) precursor, *S*-methylmethionine (SMM), specification is 5 µg DMS equivalents/g of malt. The *R* value for SMM analysis in malt (ASBC method Malt-14) is in the region of 13.5–15.2%. In other words, if you have a mean value of 5 µg/g, then, taking 15% as *R*, values within 0.75 µg/g of the mean are sufficiently precise. Let us say that the maltsters put a value of 4.35 µg/g on the analysis sheet, that being the number that they measured. The good folk at the brewery might say, "Hang on, that's too low," but they would be wrong, because it is within the *R* value. Indeed, any measurement between 4.25 and 5.75 is within the ballpark.

Figure 4-5 shows three different types of measurement patterns that can be encountered. The blue line shows a parameter than can be measured extremely precisely. The *r* and *R* values are going to be very small. An example might be the measurement of alcohol by gas chromatography. It is possible to be very precise, so *r* and *R* values are very low. In some parts of the world, it is essential to be precise on this parameter. For example, in the United Kingdom, a duty is levied incrementally for every 0.1% increase in ABV. Accordingly, you have to be very precise (and accurate!).

The measurement of SMM might be an example of the green line. There are several stages to the method—the milling of the sample, heating it with sodium hydroxide, and then the gas chromatography itself. More stages present more opportunity for inconsistency. For example, is all the SMM being dissolved from the malt, and what are the losses in transfer stages within the procedure? By definition, it cannot be as precise as simply taking a tiny amount of beer and firing it onto a gas chromatograph.

All that remains to be said is that if for any given parameter the same plot is obtained time after time after time around the same mean, then that parameter is under control. It is only when a process deviates to a greater extent than this, leading to a shift in the mean or an increase in the variability, that the process is out of control.

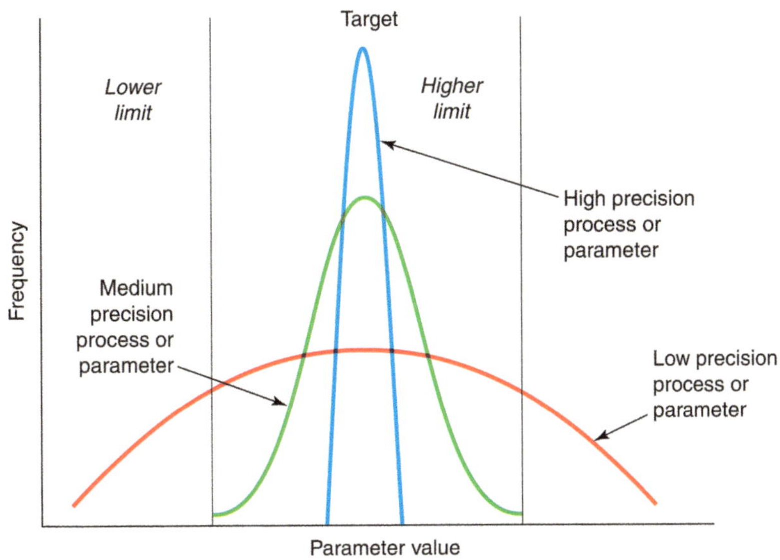

Fig. 4-5. Different levels of precision. (Reproduced, by permission, from C. W. Bamforth, 2002, Standards of Brewing: A Practical Approach to Consistency and Excellence, Brewers Publications, Boulder, CO)

Assuring Quality

Chapter 5

So we have the methods and we can appreciate whether or not they are reliable. When is the best time to apply them? The answer is "as soon as possible." By that I mean it is better to make measurements that assure you that things are going to go well rather than checking after the event that the beer is okay to go out of the door.

We can classify our measurements into three categories:

1. "Ahead of the game," to ensure that what follows goes well. In particular, I am thinking about analysis of raw materials.
2. "In real time," to allow the process to be tweaked "on the run." In other words, measurements made during the process allow the process stream to be in specification.
3. "After the event," to ensure nothing out of specification goes to trade and to allow rectification measures to be implemented. At this point it is, of course, measures on the beer itself.

Let us consider an example that shows why the earlier a measurement is made, the better things are and the less waste is incurred. Imagine that you diligently check the beer for its tendency to gush. Nightmare: You have a gusher. Bottle after bottle. Thank goodness you ran the test. It would have been horrendous had it gone to trade. That is the good news. The bad news is you are an idiot. Had you checked the malt, you would have found that it contained deoxynivalenol, a marker for strains of the fungus *Fusarium*. As we saw in *Foam*, the first volume in this series, *Fusarium* also produces hydrophobin, a major cause of gushing. So you have gone ahead and wasted time and resources using a raw material that should never have been used in the first place.

There are plenty of other checks on raw materials that allow you to proceed with them with confidence. Tasting the water to ensure it contains no taints is a classic example.

The real-time measurements include monitoring things such as specific gravity, temperature, wort clarity, and yeast count—in short, any simple test that allows you to keep the bucking bronco of a brew in harness.

Of course, the more sophisticated operations use in-line analysis with automated feedback control. An example I am fond of is the in-line monitoring of yeast count. Chris Boulton (part of David Quain's outstanding fermentation team) and I as research manager (Fig. 5-1) were the first people to bring the Aber biomass probe into the industry when we were at Bass Brewers. The delineation of responsibilities was quite simple: Chris was the brains and I drove the car from Burton to Aberystwyth in Wales, where Douglas Kell (Fig. 5-2) developed the concept and Bob Todd realized it at Aber Instruments. (By the way, because of the fact that this was the town where the device was developed, the pronunciation of "Aber" rhymes with "jabber" and not how folks like to pronounce it Stateside with a hard A.) The principle is that living yeast stores a charge and therefore acts as a capacitor. By measuring the capacitance of a suspension of yeast, you can quantify the total number of living cells therein (Fig. 5-3). So now all over the world these probes are inserted into lines, allowing for automated pumping of exactly the correct number of yeast cells to ensure a consistent fermentation (assuming, of course, you are also ensuring that all the other relevant parameters are under control, notably wort strength, wort clarity, temperature, oxygen level, zinc, and fermenter geography).

With regard to the final product, of course, the list of potential measurements, if not endless, can certainly be quite comprehensive. In Chapter 7, I discuss what is and what is probably not worth measuring.

First, though, let us dwell a wee while on how we should keep an eye (or both eyes) on what is trending in our materials, process, and products. It makes huge sense to plot out our performance. My favored way is illustrated in Figure 5-4. You will see on this diagram the target value and, above it and below it, two other lines: the AQL and

Chapter 5

Assuring Quality

Fig. 5-1. Good colleagues. Top, a quiz-winning team from the research department of Bass Brewers, Burton-on-Trent, circa 1987. Left to right: Chris Boulton, Geoff Large, Charlie Bamforth, and Paul Hegarty. Bottom, Charlie Bamforth, Chris Boulton, and David Quain in Kegworth, U.K., 2018. (Top: Courtesy C. W. Bamforth—© ASBC. Bottom: Courtesy D. E. Quain—Reproduced by permission)

Fig. 5-2. Douglas Kell, then and now. (Courtesy Douglas Kell—Reproduced by permission)

the RQL. AQL stands for "agreed quality limit" or "acceptable quality limit," while RQL is "reject quality limit." The AQL lines are set at two standard deviations (σ) above and below the target, while the RQL lines are three standard deviations above and below the target. In Chapter 3, we went into detail on the normal distribution and the fact that we expect 95% of our values to be within two standard deviations and 99% within three standard deviations. Six sigma, remember. And so for any parameter, if our value is within this six sigma range, it is not rejectable, but we prefer it to be within two standard deviations above or below the mean because we know (think about the Youden plot) that any value inside this range is very much "in specification."

As we said in Chapter 3, there are some parameters for which those tramlines are going to be very close to the mean because they are values that can be controlled very tightly. Alcohol is a great example. Total in-pack oxygen should be another. For other parameters, the tramlines are going to be farther apart because they are less precisely measured.

I would expect to see plots of this type for every significant parameter measured in raw materials, process, or final package.

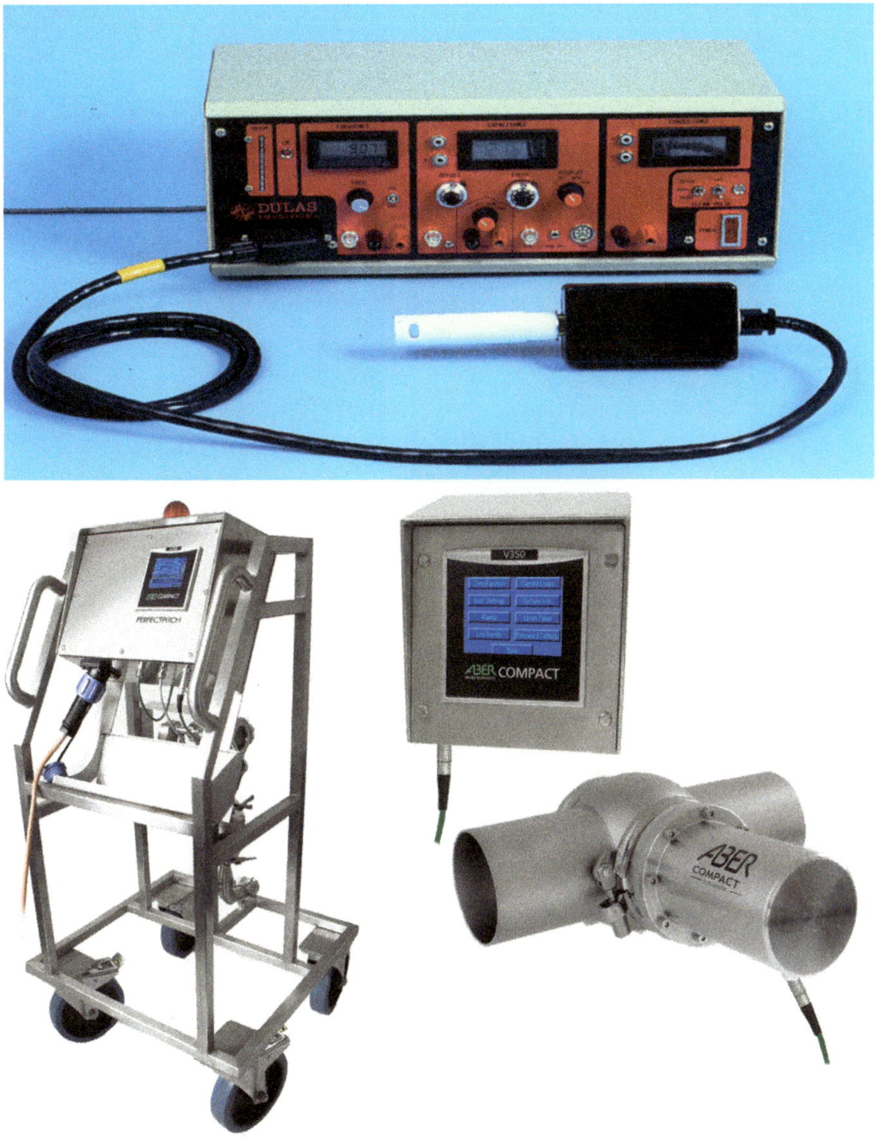

Fig. 5-3. Aber biomass instruments through the ages. Top, the first prototype. Bottom, the current Perfect Pitch system. (Courtesy Aber Instruments Ltd.—Reproduced by permission)

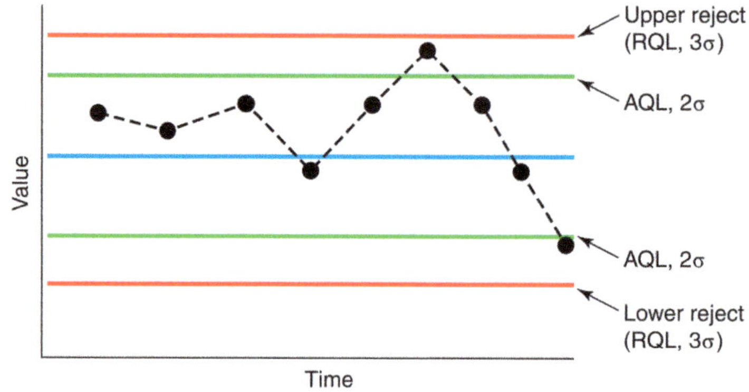

Fig. 5-4. Control diagram with warning and action lines. AQL = agreed quality limit or acceptable quality limit, RQL = reject quality limit, and σ = standard deviation. (Reproduced, by permission, from C. W. Bamforth, 2002, Standards of Brewing: A Practical Approach to Consistency and Excellence, Brewers Publications, Boulder, CO)

Quality Assurance Versus Quality Control

To illustrate the difference between quality assurance (QA) and quality control (QC), let me use an example that plagued my life from 1988 to early 1991. Bass had exiled me (akin to how the Soviets punished their dissidents with a vacation in Siberia) to the Preston Brook Brewery (Fig. 5-5). They told me that I had to have the smooth edges of the "ivory tower" knocked off me. In reality, it was a great thing to have happen, and I learned a great deal. I was particularly fortunate to pick up the QA manager baton from Dr. Harry White.

One thing I quickly learned was how not to design a brewery. Breweries are best when they are vertical, taking advantage of gravity for liquid flow. Engineers designed this one with little passion for brewing, and it was built with long pipe runs, including the line that delivered the freshly filtered beer from one end of the site to the bright beer tanks many yards distant. In came the oxygen, out went the carbon dioxide. It was the norm for us to be out of specification for these gases 80% of the time. The corporate target was 80% "right first time" for gases. Therefore, eight times out of 10, we had to couple on a carbon dioxide supply to purge out oxygen and to restore the carbon dioxide to specification. Invariably, we lost foam-stabilizing potential and produced foam, which duly collapsed to leave bits in the beer.

Fig. 5-5. **Preston Brook Brewery. Top left, the brewery, and top right, the kettles used for boiling the wort. Bottom, Harry White greeting his successor as quality assurance manager, Charlie Bamforth, in 1988. (Above, left and right: Courtesy The Brewery History Society—Reproduced by permission. Bottom: Courtesy C. W. Bamforth—© ASBC)**

The QC was working—all that repeated checking for gas content ensured that the beer left the brewery in specification. However, there was little quality assurance about any of this! All we were pretty much assured of was failing to hit the gas targets and a routine knuckle wrap from headquarters.

QA would have been to design a brewery that delivered beer in specification "right the first time" (RFT). That is the overarching mission of QA: to deliver the right product with minimum failure and rework all the way through the operation.

Now do not get me wrong. Bass was a marvelous company. All brewers have disasters. I guess they do not all have folks like me shouting their mouths off! Bass's management was very much determined to apply the finest quality tools and techniques. Thus, we embraced total quality management (TQM). Everyone has a customer. If you consider beer, then the grower of barley has as her customer the maltster. The maltster's customer is the person responsible for malt intake in the brewery, who may or may not be the chap running the brewhouse. In turn, that individual's customer is the brewer in charge of fermentation and cellaring. And so on, all the way down the line. If everyone is hitting the quality targets for his or her role in the operation, then at the end of the day the ultimate customer, the person buying the beer, should be purchasing with confidence.

Therefore, QC is a reactive device—making measurements from start to finish ensuring that everything is within specification.

QA is proactive. It is putting in place processes and procedures that assure that a quality operation is achieved: auditing of suppliers; plant design; automated control systems; the maintenance of standard operating procedures, perhaps embraced in accreditation by the International Organization for Standardization (or even addressing the principles of such a standard if not formally getting the accreditation); training of staff; ring analyses. The list goes on—and we will further illustrate the principles involved in the next chapter.

Critical Stuff

Chapter 6

Let us think about dimethyl sulfide (DMS), which we encountered in *Flavor,* volume 2 of this series. You will recall that brewers either love it or hate it but that it features as a component of the aroma of many lagers. The important thing we have established (I hope) in this series of books is that we are striving for consistency and that one person's joy (e.g., DMS, cloudiness) is another person's nightmare. We just have to control things. That is all.

So think back to DMS. I said that a key precursor is *S*-methylmethionine (SMM) in malt. The more SMM, the greater the potential for DMS to get into beer. The next thing I pointed out is that SMM is converted to DMS at the boiling stage and that if there is vigor at that stage, the DMS is driven away. I also said that the whirlpool is hot enough to degrade SMM, but the swirling there is so gentle that any DMS stays behind. So straightaway we can realize that if we do want DMS in our beer we should have (a) a substantial amount of SMM in the malt, (b) not too long a boil, and (c) a long whirlpool stand. If we do not want DMS in the beer, we should have the opposite of these three.

Now if you read, mark, and inwardly digest (as an Episcopalian I know that to be from the Collect for the Second Sunday in Advent) what I wrote in *Flavor,* you will realize that DMS control is somewhat more complex than this, but those three requirements referred to in the previous paragraph are critical control points (CCP). CCPs can be identified for every aspect of process and product, and I would hope that by now you can go back to volumes 1–4 and work out for yourself where the CCPs are for the various quality attributes. It is at these places that the greatest focus needs to be.

Let me illustrate this some more by looking at DMS. I pointed out in *Flavor* that there is very little conversion of SMM to DMS in most infusion mashing operations and during lautering. Therefore, these are not CCPs for SMM. Very little happens to SMM or DMS during cold conditioning. Not a CCP. But I did point out that DMS can be produced by wort spoilage bacteria, and I did point out that DMS is lost to differing extents with the carbon dioxide coming off a fermenter. Therefore, we have to ensure that we do not have bacteria crawling in our brewhouse (for most beers!). In addition, we should specify what fermenter we are going to use for our beer. However, you will realize that these are *general* requirements for all beers and for all quality criteria. Therefore, they would be read but not taken as a specific list of CCPs for the control of DMS.

Hazard Analysis Critical Control Points (HACCP)

When folks hear the words "critical control points," it straightaway puts them in mind of Hazard Analysis Critical Control Points (HACCP). Rightly so, in many ways, because this is all about safety, and without safety we have nothing. Furthermore, it may be a legal requirement for you (see Chapter 2).

HACCP is a management system designed to assure the safety of food products (www.fda.gov/food/guidanceregulation/haccp). In many ways, it is analogous to ISO 9001, in that it lays down a series of requirements that must be met. You will recognize the first one! Commitment from the top!

Prerequisites for implementing HACCP are

- A policy declaring the company's commitment to making safe products and all staff being familiar with it.
- Ensuring that there is no threat from surrounding industries and others who might present a risk, either accidental or malicious.

- Identification of process stages presenting the highest risk (for example, open vessels and packaging areas, including risks from glass, and there being a register for the latter).
- Fitness for purpose in buildings and equipment, ensuring that they are properly cleaned and maintained. Issues include toilets, hand washing, and pest control.
- Rigor in all suppliers of materials, with checking procedures at receipt. For example, vehicles transporting raw materials should not be employed to carry undesirable and incompatible entities.
- Proper training programs for all employees.
- A robust product-recall system.

Successful implementation of HACCP includes the following:

- Defining exactly what needs to be embraced throughout the facility from raw materials in to products and co-products out.
- Establishment of teams whose job it is to ensure the success of the program. They will generate a process flow chart covering everything that is within the scope of HACCP. In this, they will identify all of the potential hazards and establish appropriate control measures to ensure that problems do not arise. They will address both chemical and microbiological risks (for example, undesirable materials present in malt such as aflatoxins, risks of cleaning agents entering the process stream, risks of undesirable organisms populating cooling towers, and risk of cooling liquids entering wort and beer) and physical risks (for example, broken glass or filling tubes in bottles). Through this thorough evaluation, they will identify CCPs, in other words, the stages where controls must be introduced to minimize hazards. The team will identify what the critical limits are to be at each CCP, that is, the values that represent acceptability or not. Furthermore, they will establish what records need to be kept and what corrective actions are necessary if a control parameter is not satisfied.
- Thorough training for all personnel and a system for ensuring that all of the documentation is kept up to date in the event of changes in raw materials, equipment, or declared process procedures.

Fig. 6-1. Gordon Jackson. (Courtesy Gordon Jackson—Reproduced by permission)

Perhaps an example will serve to illustrate, and I turn to a former colleague, Gordon Jackson[1] (Fig. 6-1), for this. It is shown in Table 6-1. Here is just one of the many components that you would expect to see in the entire HACCP document. It concerns avoiding the migration of coolant from the cooling side of a bright beer chiller into the beer. The simple quality assurance expedient is to make sure that the pressure on the beer side is higher than that on the coolant side, so if there is a hole, it will be a matter of beer getting into the glycol and not the vastly worse situation of glycol getting into the beer. Naturally, we do not want either! The quality control elements concern checking the pressures so that we can ensure that the *x*-bar differential is maintained. However, periodic quality control checks would also include analysis of coolant in beer (and beer in coolant) as well as a physical examination of the integrity of the cooler plates.

---

[1] Gordon Jackson has been a remarkable servant of a single organization at the heart of brewing research and investigation in Surrey, England. As I write, he is still there, having joined them in 1970. When he was appointed, they were the Brewing Industry Research Foundation. It has successively been the Brewing Research Foundation, Brewing Research Foundation International, Brewing Research International, and most recently Campden BRI. He displays true dedication, and if I were not dedicating this series of books to family members, I would readily dedicate this volume to Gordon.

Table 6-1.

Example of a Critical Control Point[a]

Process Step	Hazard and Cause	Control Measure	Critical Limits	Monitoring	Corrective Action
Chill beer during transfer from bright beer tank to filler bowl	Chemical contamination from secondary refrigerant caused by leaking heat exchanger	Product pressure higher than coolant pressure during beer transfer	Pressure differential = x-bar	Check coolant inlet pressure and product outlet pressure. Frequency: once per hour Responsibility: A. Person	Stop beer forward flow. Examine heat exchanger and repair. Responsibility: A. Person Isolate product produced since last check and analyze for presence of secondary coolant. Responsibility: A. Person

[a] Adapted from G. Jackson, 2006, Quality assurance in brewing, pages 358-371 in: Brewing Technologies, C. W. Bamforth, ed., Woodhead Publishing, Boston. Courtesy C. W. Bamforth—© ASBC.

Hazard Analysis and Risk-Based Preventive Controls (HARPC)

Hazard Analysis and Risk-Based Preventive Controls (HARPC) is a necessary upgrade to HACCP. The U.S. Department of Agriculture and the Food and Drug Administration (FDA) insist on HACCP for only meat, seafood, and juice products. HARPC goes beyond CCPs and is broadly mandated by the FDA. It enforces preventive controls to identify potential risks or threats and the implementation of appropriate proactive corrective measures to prevent contamination. HARPC involves the development of a preventive controls plan based on FDA regulations, standards, and guidance documents (Table 6-2).

Table 6-2.

Comparison of Hazard Analysis Critical Control Points (HACCP) and Hazard Analysis and Risk-Based Preventive Controls (HARPC)[a]

Key Comparison Points	HACCP	HARPC
1. Is the preventive approach based on a standard, guideline, or set of laws?	Based on a guideline recommended by CODEX and the National Advisory Committee on Microbiological Criteria for Foods	Based on the Food Safety Modernization Act (FSMA) and, principally, the final rule for Preventive Controls for Human Food
2. What food safety risks are considered in using the preventive approach?	Conventional—biological, chemical, and physical	Beyond the conventional risks for actual and potential food safety hazards
3. What is the goal of the preventive approach?	To prevent, eliminate, (or) reduce hazards to a safe level (in that order)	Controls that prevent or significantly minimize "known or reasonably foreseeable" risks
4. Who is primarily responsible for the development and maintenance of the preventive plan?	A competent HACCP coordinator with assistance from a multi-disciplinary team	Trained Preventive Controls Qualified Individual, as described in FSMA
5. At what frequency is the preventive plan reviewed by the facility?	At least once a year, or when required	At least once in 3 years, or when required
6. The plan is mandatory for what type of establishments?	Establishments mandated by the U.S. Food and Drug Administration (FDA) and the U.S. Department of Agriculture (USDA), or when required for certification purposes	All establishments along the food-supply chain that serve U.S. consumers, unless exempted
7. What types of establishments are excluded or exempted from the plan?	Unless mandated or required for certification, HACCP is voluntary, and Good Manufacturing Practices (GMPs) are mandatory	Exemption list is provided by the FDA, but this does not exempt facilities from following at least current GMPs
8. Who is the interested party here? For whom is the plan intended?	Stakeholders, auditors, inspectors, and customers	FDA
9. What is the documented approach for making the preventive plan?	12 steps of HACCP (includes seven principles) (see text)	Seven steps of developing a HARPC plan (see text)

[a] Adapted from Food Safety Magazine, Sept. 19, 2017, with permission of the publishers. © 2019 by The Target Group. https://www.foodsafetymagazine.com

The assessment of food safety hazards is broader within a HARPC plan, covering these risks:

- biological, physical, chemical, and radiological hazards
- natural toxins, pesticides, drug residues, decomposed material, parasites, allergens, and unapproved food and color additives
- naturally occurring hazards
- unintentionally introduced hazards
- intentionally introduced hazards, including acts of terrorism

All hazards "known or reasonably foreseeable" must be addressed by a team of qualified and properly trained people as defined in the Food Safety Modernization Act, which was enacted in 2011. The HARPC plan must be reviewed, at a minimum, once every 3 years and earlier if any substantial change has been made in the facility that has bearing.

The steps for HARPC are

1. Assess the hazards.
2. Institute preventive controls, including sanitation procedures, training of staff, environment monitoring, and supplier verification.
3. Monitor effectiveness of the controls. Not all controls are measurable by critical limit numbers, but these preventive controls can be evaluated on a routine basis.
4. Establish corrective action measures; that is, identify and correct problematic features.
5. Establish verification measures to confirm that the facility is consistently meeting the food safety standards.
6. Follow required record-keeping procedures.
7. Reanalyze the plan every 3 years or as needed (see above).

HARPC differs from HACCP in that science- or risk-based preventive controls are required as opposed to CCPs. Setting up critical limits is not required under HARPC, but the establishment of preventive controls to minimize risks must be supported by reliable scientific data or literature.

The Cost of Quality

Okay, we are forearmed with our systems, such as HACCP and HARPC. We have our methods and our standard operating procedures. We know what we want to measure and how often (of which more later). Moreover, we are rather pleased with our brewery. What else could we need?

Let me paint a picture. There is a new managing director at the helm. He is a graduate of some fancy-dandy business school at an Ivy League university. He was previously with a company that led the world in the manufacture of ball bearings. He took his new job not because of any love for beer (in fact, he is partial to Prosecco) but for other professional reasons having to do with career enhancement. And there is you, a relative pawn in the game, heading up the beer cellar operations in one of the company's breweries. Your nightmare is oxygen pickup. Day in and day out, your beer arrives in the bright beer tank (BBT) with an oxygen level of 0.06 mg/L, whereas the specification is 0.05 mg/L. El Supremo deigns to make a visit to this outpost of his empire, and you meet him for the first time. He asks whether you have any concerns—not expecting anything quite so specific as when you tell him that there is too much oxygen in the beer. He asks what the problem with that is, and you say that the more oxygen, the more rapidly the beer will stale. He asks how much should there be, and you say less than 0.05 ppm. And how much do you find? You say 0.06 ppm. He is not impressed.

"That is a tiny difference," he says.

You say, "The simple reality is that the more oxygen, the more rapidly the beer will stale."

"Okay," he says, "but you can correct it, can't you?" with a "get on with it" tone.

"We certainly can," you say. "What we do is measure the oxygen in the beer when it arrives in the BBT. If there is too much, then we get Jenny, my assistant, to couple up the carbon dioxide pipe and we purge off the oxygen. Then we get the lab to analyze the beer again. If we just take the employee costs for doing this, then each oxygen sampling and analysis costs about $25 and Jenny's efforts come in at something similar. Each BBT holds 1,000 barrels, and last year we shifted 1 million barrels of beer. That means we filled the BBTs 1,000 times. We had to do the correction 80% of the time, so that means this problem cost us $40,000." You now have the bean counter's attention.

We are talking the *cost of quality.* It can easily be the difference between success and failure in a company. We can divide the cost of quality into

- the cost of internal failure
- the cost of external failure
- the cost of appraisal
- the cost of prevention

We have already come across a couple of examples of internal failure earlier: the real-world one of too much oxygen that I mentioned in the previous chapter (the example given in this chapter is a hypothetical illustration of this) and the gushing example. Anything that goes wrong and needs correction is an example of failure. Prolonged lautering times mean fewer brews per day, failure; stuck fermentations, failure; packaged beer with a metallic off-flavor, failure. You can readily convert these disasters into monetary terms analogous to what I did earlier: the cost of doing things again (rework), the reinspection costs, the cost of having to scrap a product stream, and the fiscal (as well as emotional) cost of blame seeking.

If internal failure is bad, then external failure is heaps worse. The worst beer example I know of comes from the United Kingdom back in the 1990s. A very, very prominent brand was found to contain glycol coolant (if only Gordon Jackson had written his chapter at that point). The company did the decent thing (many would not have—the amount of coolant contamination was small and would not have killed anyone) and announced the problem, recalling the beer from trade. There was the cost of this physical pullback, the cost of replacement, and legal and liability costs, but, above all, there was the loss of reputation, manifesting itself in a plunge in the price of the company stock. Not only automobile companies have public reckonings.

Appraisal costs include the price of analyses throughout the process, from raw materials to end product inspection and test, internal and external auditing and approving—in short, the day-to-day cost of quality control.

Which leaves prevention costs, which we might refer to as the cost of quality assurance, that is, the expenditure on all of the systems that will minimize failure. It is the cost of equipment (such as in-line analysis), the cost of upgrading parts of an operation that are not fit for purpose (think the oxygen example), the cost of documentation and systems (such as HARPC), and the cost of training and education (buying and circulating this book would be a fabulous example!).

Increased spending on prevention should decrease the need for ill-considered and wasteful analysis (especially reanalysis), but even more so, it should ensure far less internal and external failure. The result is actually a financial saving.

Setting Up a Quality Program

Chapter 7

Armed with the basics covered in the first six chapters, we are in a position to consider how we should approach establishing our quality system. Let us suppose that we have been newly appointed the quality assurance (QA) manager for a brand new brewery (and we will not dwell on the size of the company; let us assume the same basic principles that we must address are common to all breweries). We are charged with the task of preparing the list of what is essential. For simplicity, we are assuming that the brewery is perfectly designed, e.g., cleanable.

We first need to make quite clear a few overriding principles that everyone needs to buy into.

First, nobody should be looking at the QA manager and his team as some form of irritants who are there to make life difficult for those responsible for brewing the beer. Quality is everybody's responsibility. Therefore, whatever is set up needs to reduce confrontation to the greatest extent possible. Thus, it is imperative that all employees have a clear understanding of the impact of their actions and performance on quality (in other words, they should be educated) and furthermore that, to the greatest extent possible, those producing the beer should be involved in making measurements.

Second, the protocols should match the market for that beer. For example, if our brewery is making beer that is going to be served from a tank in the cellar to the one outlet for our beer, namely, the bar upstairs, then the process and measurement regimes will be somewhat different from those for a brewery that is intent on conquering the world with its beer and shipping it across the globe. In the latter instance, there is going to be much more interest in procedures and analyses that relate to long-term stability of the beer.

The third guideline is what I call the "master chef" approach. There needs to be a hands-on approach to the raw materials and processes. As we saw in Chapter 5, there are three types of measurement (those conducted before, during, and after brewing), and the more sophisticated tests are the ones applied earlier in the process, followed by those used later in the process. Just as a chef is carefully selecting the ingredients and scrutinizing events as the kitchen activities unfold, most notably through taste, so it is with beer. The focus should be primarily on proactive techniques and not on reactive ones.

Fourth, if you are not going to use a number, then do not ask for it or generate it. An example might help. When I was a new QA manager, I scrutinized the list of analyses that we were providing to the brewers and saw that there was a line for total soluble protein (nitrogen) (TSN) in wort. I stopped making that measurement. The head brewer confronted me.

"Where is the TSN value?" he asked.

I told him that we had ceased to generate that number.

"But I want it," he insisted.

"Why?" I retorted.

"Because I want to make sure that it is in specification."

"And when wasn't it in specification?" I asked.

"Well, never," he admitted.

I went on, "And if it had not been in specification, what would you have done differently?"

He was at a loss to respond.

"It doesn't tell you anything about foam potential. It doesn't tell you anything about haze potential. It doesn't tell you anything."

In the end, I compromised and said that we would freeze a sample of the wort and if ever he needed the TSN, I would thaw it out and do the test (though I am sure the freezing would have screwed up the figure anyway!). He never did come back with the request for us to do this.

Fifth, pass the buck. In this day and age, there really is no justification for a brewer to endlessly analyze raw materials, for instance, other than sensorially. A delivery of malt should come with an analysis sheet, confirming that the batch is within specification. Provided that there is a good relationship between the maltster and the brewer, there is every reason to expect that the malt is on target. And the same is true for the other raw materials. Perhaps the biggest risk is with the water, insofar as the supplier may be switching between reservoirs that hold water with somewhat different characteristics. Remember, though, that above all things, the number one criterion we should be applying is taste and we are going to have the QA expedient of filtering the water through carbon. We expect the water company to send us the detailed ionic analysis and to verify that the water meets the National Primary Drinking Water Regulations. Our good relationship with suppliers involves some form of routine auditing of that supplier (perhaps short of the sliced bread test[1]), which for a small company might mean being part of a consortium of similar-sized entities that commission a third party to do the auditing on their behalf. Equally, it might mean periodic spot-checking of the material by analysis in one's own lab or through a third party.

Sixth (and certainly not least), insist that as QA manager you do not report to the person whose day-to-day responsibility is brewing the beer and/or packaging it. You need to report to someone other than that person. If that does not happen, production expediency will always take precedence over quality. Let me give you an example from my time as a QA manager.

I arrived one morning and, as usual, made my way to the packaging hall to check the log, i.e., the book in which the quality staff on the night shift had recorded key happenings. Straightway I noticed that they had held a beer, in other words, put an order on it that stipulated that it could not leave the brewery pending approval by a higher authority. The problem was that it was over specification for total in-pack oxygen. In those far-off days, I believe the spec was 0.1 ppm, and the beer was reading 0.11 ppm. Within minutes, the managing director was on the phone. Now, he was a phlegmatic individual, ideally suited to the management of some incendiary characters employed on site. He quietly asked me to make my way to his office, where I was

[1] Representatives of a very large U.S. brewing company would arrive to audit the malthouse and present a piece of sliced bread to the maltster. The latter was invited to wipe it on the inside of the steep tank and enjoy it for breakfast. The extent of enjoyment was naturally in proportion to the cleanliness of the vessel.

confronted by a very angry packaging manager and a fuming warehouse manager.

"What's happened, Charlie?" asked Mike Myers (not that one).

"We've held this beer."

"Why?" he asked evenly.

"It's got too much oxygen in it."

"So why should we worry about that?" said the managing director.

"Because it will stale more quickly."

"I see," Mike replied. Turning to the warehouse manager, he asked, "And what's your take?"

"We need to ship this beer. The trade is out of it and we need to get the shelves stocked urgently."

Mike turned back to me.

"So what do you say, Charlie?"

"Okay, I'll release it."

"Why the sudden change of heart?" Mike queried.

"Well, if it's going to be bought and consumed soon, then staling is not going to be an issue."

Everyone calmed down. At least before my next statement, which was, "But the packaging folk need to get their act together and make sure it doesn't happen again."

If I were reporting to the production folks, that discussion would never have taken place and the beer would have been shipped regardless. The simple expediency of my reporting to the managing director and having the power to hold out-of-specification beer allowed a reasoned debate to take place in the interests of the best balance between output and quality.

A Baseline Approach

Employing these principles and armed with the tools discussed in the previous chapters, let us lay down the simplest framework we might use to meet the fundamental goal, namely, the production of the same beer time after time after time.

At the outset, it must be stressed that representative sampling has to be done. Whatever is measured needs to be representative of the whole. For instance, say you have malt arriving by truck. Only by taking samples from different depths on the truck can you be confident that there are not different batches in that consignment. Now, as I point out elsewhere, a degree of trust must exist between user and supplier so that we are confident that this situation will never arise—woe betide the supplier if it does! For those in a trusting relationship, then, this should not be a consideration. Say, though, you are brewing on a much smaller scale, getting your malt in sacks. Representative sampling now becomes each bag of malt. Similar considerations apply to other ingredients. For the final package, one needs to consider the frequency with which samples are taken for checking. Think, for example, of package checks in canning (for example, seam analysis and contents). One can taken at random is insufficient. Cans drawn from the line at intervals must be checked.

Malt

A guideline for the line items that you should expect on a malt specification and analysis sheet is shown in Table 7-1. The golden rule is to have on that sheet only items that are relevant and needed. The more analyses expected of the maltster, the more challenging you are making life for him or her and the greater the likelihood that he or she will be blending to meet specifications, which in itself is not a good idea.

Note that this table is not comprehensive. Some will look for wort viscosity and total β-glucan, but these values should correlate with friability. Some look for coarse–fine difference, but that is a fairly imprecise measure of modification and should correlate with friability.

Other items that really are only for the purist include diastatic power, dextrinizing units, β-glucanase, total protein, soluble nitrogen ratio (Kolbach index, or S/T value), and fermentability. They represent parameters that are for the most part informed by the key indices shown in the table.

Another line item could be the premature yeast flocculation factor. A simple + or – for this might be expected.

Two food safety risks are nitrosamines and deoxynivalenol and may be stipulated as line items by some. The former is highly unlikely to be present because indirect firing is used in modern kilns. The latter is produced by *Fusarium* spp. A reddish coloration produced by the fungus is readily apparent on visual inspection. This is highly unlikely to occur in North America.

Table 7-1.

Key Malt Specifications[a]

Parameter	Typical Value	Justification
Pale malt[b]		
Variety	Not applicable	Brewers "know" the varieties they prefer. There are some who believe it is relevant for beer flavor.
Absence of taints	No numerical value; just a confirmation that the malt smells, tastes, and looks good	Check that there is no sign of contamination or taints.
Moisture	Ale, 3.5% maximum Lager, 4.5% maximum	Guideline to stability of the malt and severity of kilning
Fine grind extract (dry basis)	Ale, 82% minimum Lager, 80% minimum	Indicator of potential extract derivable in the mash
Friability	Ale, >85% Lager, >80%	Measure of modification of the malt and extent to which problematic materials may remain that will retard wort separation
Homogeneity	>95%	Less than 5% grossly undermodified grain in the sample. Undermodified grain yields very poor extract and delivers problematic materials, like β-glucans, into the mash and wort.
Color (SRM)	Ale, 2.5–3.5 Lager, 1.5–2.0	Indicates appropriate kilning
Free amino nitrogen	100–150 mg/L for a wort of 10°Plato	Too high a value is of concern from a microbiological and flavor-stability perspective. Too low a value predicts a fermentability problem.
Dimethyl sulfide (DMS) precursor (*S*-methyl-methionine)	Lager malts, from a low of 1–2 up to a high of 15 μg DMS equivalents per gram of malt	Depends on whether DMS is desired in the lager. A value of 6.5 might lead to a DMS level in a beer of about 50 ppb, assuming a 1-h boil and a 1-h whirlpool stand

continued

[a] Courtesy C. W. Bamforth—© ASBC.

[b] The malt that forms the bulk of the grist as a source of starch and enzymes.

[c] For the most part, these malts are used to introduce certain flavor and color characteristics. There are no specific flavor substances defined, so fundamentally the key parameter analyzed is color.

Table 7-1. continued

Key Malt Specifications[a]

Parameter	Typical Value	Justification
Age of malt	>4 weeks	Some degree of storage of malt is necessary, otherwise the wort lauters poorly. I err on the side of 4 weeks minimum. Therefore, the "kilned on date" is key.
Specialty malts[c]		
Absence of taints	No numerical value; just a confirmation that the malt smells, tastes, and looks good	
Moisture	Caramel malts, 3–7% (inversely proportional to color) Chocolate malt, 5.5% Black malt, 6.5% Roasted barley, 5% Black barley, 6%	
Color (SRM)	Caramel malts, 10–120 Chocolate malt, 350–420 Black malt, 500 Roasted barley, 300 Black barley, 500	

Wheat malts and any other relatively lightly colored malts that constitute the primary source of enzymes and extract for a grist should be judged in a manner comparable to the above.

For more detailed information, see the chapter by N. Davies in Bamforth (2016).

Your job (or somebody else's job) in the brewery is to ensure that the parameters listed on the data sheet are within specification and that you have physically inspected the malt to ensure that it looks and tastes as expected and includes no undesirable materials such as metal and rocks.

Adjuncts

The same considerations apply to adjuncts. Ensure that the analysis sheet marries with the spec sheet (Table 7-2) and that due diligence is done on the visual/olfactory/taste physical inspection.

Table 7-2.

Examples of Key Adjunct Specifications[a]

Parameter	Typical Value	Justification
Absence of taints	No numerical value; just a confirmation that the material smells, tastes, and looks good	Checking that there is no sign of contamination or taints
Moisture (%)	Torrefied barley, 6.0 Flaked barley, 9.0 Torrefied wheat, 8.5 Wheat, 12 Rice, 11 Corn, 12	Guideline to stability of the adjunct and that its production processing has been appropriate
Fine grind extract (% dry basis)	Torrefied barley, 78 Flaked barley, 75 Torrefied wheat, 79 Wheat, 91 Rice, 96 Corn, 93	Indicator of potential extract derivable from the adjunct
Color (SRM)	Torrefied barley, 2.0 Flaked barley, 2.0 Torrefied wheat, 1.5 Wheat, 2.0 Rice, 0.7 Corn, 0.7	Indicates appropriate processing

[a] Courtesy C. W. Bamforth—© ASBC.

Sugars and syrups are, for the most part, checked for the absence of taints, overall strength (which may be stated in Brix), and distribution of fermentable sugars. Most are of low color, but color may need to be checked for some, e.g., darker candi sugars.

For more detailed information, see the chapter by G. Stewart in Bamforth (2016).

Water

Naturally, we are in the same situation with water. Here, we have a legal expedient in the form of water regulations. In the United States, these are the National Primary Drinking Water Regulations (https://www.epa.gov/ground-water-and-drinking-water/national-primary-drinking-water-regulations#content), so the supplier has an *a priori* requirement to ensure that the water meets those demands. If the water is from our well, the supplier is us. When it comes to the ions that brewers fuss about as influencing their process or product, you will not find them in the regulations because there is no aspect of the law that dictates their levels. That is a matter of terroir, if you will, and the terrain from which the water is drawn. Impact what the brewer is trying to do it certainly does, so we need to have that analysis sheet to peruse so that we can compare it with the water from any given reservoir. Other than the items covered in the regulations, we might expect to be checking for any striking deviation in alkalinity, calcium, chloride, copper, iron, magnesium, manganese, silicate, sodium, sulfate, total dissolved solids, and turbidity, for the reasons shown in Table 7-3.

Above all, it is incumbent upon us to taste the water every time we do anything to it (e.g., filter it or deaerate it). We are simply ensuring

Table 7-3.

Components of the Water Supply to Monitor Other Than Those Covered in the Regulations[a]

Parameter	Justification
Alkalinity	Influences process performance, including extraction of astringent materials at the end of lautering, through raising pH
Calcium	Lowers pH, removes oxalic acid, affects yeast behavior in suspension
Chloride	Impacts beer flavor
Copper	Destabilizes beer
Iron	Destabilizes beer, gives metallic flavor
Magnesium	Impact similar to that of calcium, though less pronounced
Manganese	Destabilizes beer
Silicate	Destabilizes beer, promotes scale buildup
Sodium	Impacts beer flavor, excess is not healthful
Sulfate	Impacts beer flavor
Total dissolved solids	Index of scaling potential
Turbidity	Index of scaling potential, potential contamination

[a] Courtesy C. W. Bamforth—© ASBC.

Box 7-1. Informal tasting

I consider the most valuable sensory work to be done is the day-to-day assessment of raw materials, process streams, and final products.

I spoke about this in volume 2 of this series, *Flavor,* and make no apologies for repeating what I wrote there:

When I was quality assurance manager of a brewery, the most important tasting comprised the 9 a.m. session every workday attended by my lab manager, the head brewer, his assistant brewmaster, and me. The chap who coordinated the taste work had the samples ready for us, and we tasted them sequentially. We each had a half-pint glass and stood (literally) around a bucket. Alan poured into each of our glasses a couple of ounces of each of the samples, and we would swirl, sniff, taste, and toss away the excess. Then, one by one, we announced our scores for that sample on a scale of 1 to 4.

1. Zero defects, on target; beer can go ahead.
2. Marginally off target (and we would say in what way); beer can go ahead.
3. Significantly off target (and why); beer to be held and further analyzed and a decision to be made on what to do when more information was obtained.
4. Major defects (and why); beer unacceptable and to be disposed of (and urgently put in place systems to make sure it won't happen again).

We tasted beer at the end of fermentation (and the company philosophy was that vicinal diketone control was a fermenter event); postfiltration and prepackaging; and postpackaging. In this way, we ensured that unacceptable beer did not proceed to the next stage, incurring oncost, if it was going to be identified as unacceptable down the pike. For example, how stupid would it have been for us to package that metallic-tainted beer?

Building on that point, the daily samples also included water: the water as it arrived at the brewery and water that had had something done to it, e.g., deaeration for the dilution of high-gravity brews. The rational is obvious: The vast majority of beers are largely water, and if that water has a taint, it would be stupid to use it.

For a long time, the folks at Anheuser-Busch have tasted (or at least smelled) pretty much everything from start to finish: malt and hop teas; worts at all stages; final rinsings of beechwood chips; rinse water at the end of clean-in-place systems; and so on.

Of course, we need not restrict ourselves to the senses of smell and taste. What are our eyes telling us when we inspect process items and our facilities? Foam, color, and clarity of liquids; off colors in raw material; defective labeling of bottles. The list is endless.

For more detailed information, see the chapter by W. J. Simpson in Bamforth (2016).

that it is not tainted in any way. This tasting protocol does not have to be very complex (Box 7-1).

For more detailed information, see the chapter by M. Eumann and C. Schaeberle in Bamforth (2016).

Hops

I am fond of pointing out that the most common depiction of brewers on film is of them rubbing and smelling hops, looking both for a clue that they have the right variety and checking that there is no flavor defect there, such as the cheesiness associated with deteriorating hops. Of course, most folks are not using whole-cones hops, rather pellets—and they are less fun to rub! However, the simple rule that we have already established and will continue to come back to is that we need to do a physical examination of whatever hop material we have to make sure it looks and smells like we expect it to.

The list of elements on a specification sheet for hops that demand scrutiny is not long (Table 7-4). Brewers who demand the more rigorous control of perceived bitterness and who worry about the proportions of different types of iso-α-acid in beer may also scrutinize the cohumulone content (iso-cohumulone in preisomerized material) (see *Flavor*). Also, those using hop extracts may test for gushing potential (see volume 1 in this series, *Foam*).

For more detailed information, see the chapter by T. R. Roberts in Bamforth (2016).

Table 7-4.

Key Specifications for Hops and Hop Products[a]

Parameter	Justification
Variety	Hop varieties and their derived products vary greatly in terms of aroma and resin content
Absence of taints	To ensure there is no deleterious element to the product
Alpha acid (iso-α-acid in preisomerized products)	Allows calculation of hopping rate in controlling the bitterness value of the finished beer; assumes the brewer understands the utilization rates in the brewery

[a] Courtesy C. W. Bamforth—© ASBC.

Yeast

The analytical work performed on yeast very much depends on how we are handling it. Are we buying sachets of liquid yeast that we are going to pitch in directly? Are we buying dried yeast that we are going to "reconstitute" prior to pitching? Are we simply going fermentation-to-fermentation with our yeast, perhaps ad nauseam, as per some traditional English breweries? Are we collecting yeast between fermentations and processing it, e.g., acid washing it?

The core list of what might justifiably be measured in a QA program for yeast is presented in Table 7-5. This is a baseline set of methods. The more sophisticated laboratories make measurements such as yeast vitality, but there is no universal agreement on the reliability of these methods.

For more detailed information, see the chapter by I. Russell in Bamforth (2016).

Hygiene Status

As I pointed out in volume 4 of this series, *Color and Clarity,* if you do not have a clear appreciation of hygienic processing, you should seek another career or hobby. A list of relevant investigations is in Table 7-6.

For more detailed information, see the chapters by D. Loeffler in Bamforth (2006) and A. E. Hill in Bamforth (2016).

Mill Settings

The correct mill setting is imperative if we are going to get a satisfactory yield of extract and a good wort separation process. For simplicity, I will restrict consideration to a two-roll mill–mash mixer–lauter tun operation. The gap setting must be measured regularly (perhaps weekly, depending on how frequently milling is performed). For this purpose, feeler gauges are employed (Fig. 7-1). If the gap is too narrow, overmilling and damage to the husk occur. If the gap is too wide, the average particle size distribution increases, making for less effective dissolution of extract. Another way to check that the mill settings are correct is to do a sieve analysis of the milled grist.

Table 7-5.

Key Control Parameters for Yeast[a]

Parameter	Method	Justification
Strain	Approaches summarized by Quain (2006). Most people send a sample to an expert lab for analysis. Fundamentally, is an issue of careful internal yeast handling in multiyeast facilities and faith in yeast suppliers.	Use of correct strain for a given beer; avoidance of unexpected behaviors (e.g., flocculation, changes in flavor).
Age	Microscopic counting of bud scars. Most labs do not have a powerful enough microscope, so good yeast husbandry and good record keeping are necessary: attention to numbers of successive fermentations with a given batch of yeast; optimized collection;[b] optimized interfermentation handling (e.g., conditions in yeast brink, acid washing).	Old yeast does not perform well and can lead to problems with flavor control and quality deterioration (e.g., damaged foam).
Viability	Methylene blue staining	Dead yeast does not ferment and introduces dangers from autolysis.
Absence of contaminants	Microscopic inspection for non-yeast material[c]	Bacteria and wild yeasts jeopardize quality.
Cell number	Hemocytometer (combined with methylene blue for viability); biomass probes	Part of the equation for controlling fermentation profile; more reliable than other tests for quantifying yeast.

[a] Courtesy C. W. Bamforth—© ASBC.
[b] For more detailed information, see Powell, C. D., Quain, D. E., and Smart, K. A. (2004) The impact of sedimentation on cone yeast heterogeneity. J. Am. Soc. Brew. Chem. 62:8-17.
[c] For more detailed microbiological exploration, see the chapter by I. Russell in Bamforth (2016).

Table 7-6.

Monitoring the Hygiene Status of the Process[a]

Parameter	Methods
Robustness of cleaning system, including strength of cleaning solutions, times, and temperatures	See chapter by Loeffler in Bamforth (2006)
Hygiene status of all aspects of the facility	See chapter by Hill in Bamforth (2016)
Periodic evaluation of microbiological status of raw materials and process stream	See chapter by Hill in Bamforth (2016)

[a] Courtesy C. W. Bamforth—© ASBC.

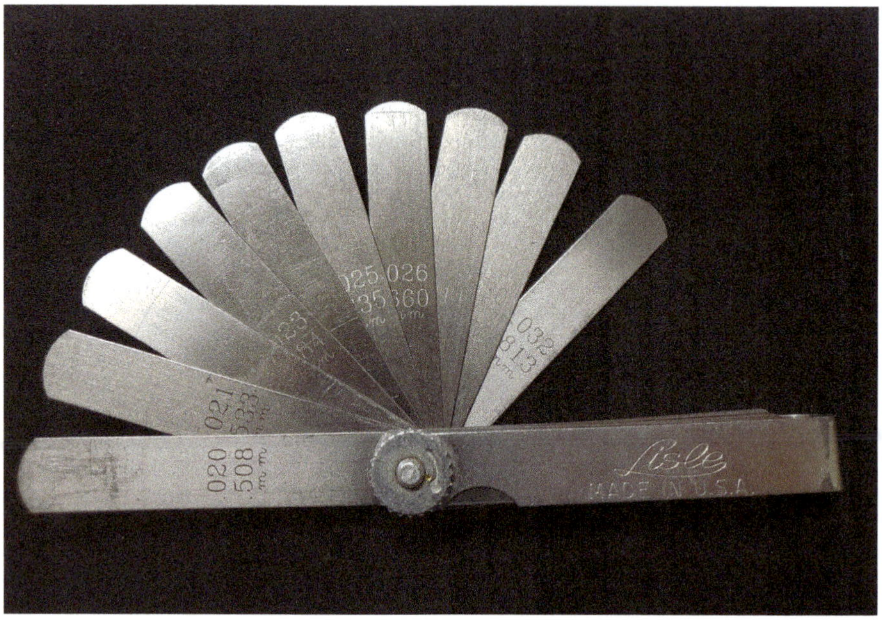

Fig. 7-1. **Feeler gauges for measuring the gaps between rolls in a mill. (Courtesy Joe Williams—Reproduced by permission)**

Mashing

There are, of course, four essential parameters that must be controlled and therefore quantified at this stage: the mass of grist materials (primarily milled malt), the volume (or mass) of water used to mix with the milled grist, the temperature of that water, and the time. Our standard operating procedures stipulate the required water-to-solid (liquor-to-grist) ratio, i.e., the thickness of the mash in relation to the strength of the wort we require to collect in the kettle prior to boiling. This is in accordance with the properties we wish to achieve in our wort in terms of not only extract but also parameters such as color, flavor, and foam potential. Examples of calculation of grist composition and the temperature of the water to be employed in mashing are given elsewhere (Bamforth, 2002).

Measurement of the amounts of grist materials and striking water depend on the sophistication of the brewery: at one extreme, load cells and metering and at the other extreme, scales and buckets.

I think most readers have an appreciation of how to measure time.

Iodine is used by many to assess the removal of starch in a mash. Remember that the blue color is produced only by the amylose component of the starch, which amounts to just one-fifth of the total starch present.

Some brewers use enzymes, e.g., prolyl endoproteinase (PEP; also called prolyl endopeptidase), either in the mash or later in the process in the fermenter in the production of gluten-reduced products. Critical quality approaches include optimal storage of these enzymes. (I have seen buckets of enzymes stored next to kettles—not a good idea.) Keep them cold and stored strictly according to supplier recommendations. There also needs to be some system of guaranteeing that the enzyme is added. This is especially critical for the use of PEP, because otherwise the resultant beer will not be in a form that can safely be consumed by the gluten-intolerant.

Lautering

Primary measures at the wort-separation stage are, of course, specific gravity, pH, and, for many people, clarity. For those insistent upon a more detailed approach, see the chapter by R. Pahl, B. Meyer, and R. Biurrun in Bamforth (2016).

Boil

The functions of the boil and the performance obtained is very much governed by achieving consistent vigor and duration, as well as ensuring consistent operations, e.g., quantities and timing of additions (hops, syrups and sugars, and kettle finings). Analyses might include specific gravity and volume before and after (enabling understanding of percent evaporation), pH, and perhaps color.

Whirlpool

The same applies in the whirlpool, with time a key consideration. For example, varying residence time leads to varying production of dimethyl sulfide from the breakdown of *S*-methylmethionine with the attendant risk of flavor change. Once again, we come across the simple expedient of doing things the same way every time as the basis of achieving a consistent product.

Cooling

Key is measurement of temperature ex-cooler to ensure that it is at the level required for fermentation of the relevant beer from that wort. It is also at this stage that we add oxygen (or air), ideally on the cold side. While this might be regulated on the basis of using a consistent rate of introduction of the gas, the more sophisticated may use an oxygen analyzer.

Fermentation

Fundamentally, it is important to use a given fermenter type for a given style of beer. For example, tall cylindroconical vessels lead to higher hydrostatic pressures that suppress certain volatiles, such as esters. The aim must be the same fermentation profile every time, as measured by specific gravity decrease and pH fall. To achieve this, parameters to be controlled are wort strength, temperature, oxygen concentration, and viable yeast count. It is also important that the wort clarity is consistent, irrespective of whether the belief set is that it should be as bright as possible or as turbid as possible. The particles nucleate carbon dioxide, causing movement in the fermenter and thereby affecting contact of yeast and wort and the rate of fermentation. A simple test is looking at a representative sample of the pitching wort in a glass container and checking that its appearance is "normal" for the operation.

Whether in a primary fermentation or a warm conditioning operation, the brewer must be mindful of vicinal diketones. We addressed issues of measuring these (and their precursors) in volume 2 of this series, *Flavor.*

Cold Conditioning

Consistent performance demands primarily the monitoring of temperature and ensuring that all beer is at that temperature, preferably the lowest temperature possible without freezing the beer. Again, clarity can be assessed by putting the aging beer into glass jars and inspecting it against the backdrop of a good light source.

Filtration and Stabilization

This visual inspection of clarity is, of course, again pertinent post-filtration. As we saw in volume 4, *Color and Clarity,* there is a range of instrumental methods for quantifying turbidity and assessing the efficacy of stabilization procedures.

Packaging

Table 7-7 provides a summary of the fundamental checks on packages. For more detailed information, see Klimovitz and Ockert (2014).

Warehousing and Storage

Critical areas here are little more than time and temperature, including for those beers that are being naturally conditioned.

The Beer

We assume that due diligence on all of the above ensures that the beer in the final package is in good shape. The relevant analyses and frequency with which they might be conducted are summarized in Table 7-8 (see also Table 7-7).

Table 7-7.

Key Control Parameters for Packaged Beer[a]

Parameter	Justification
Package properties	Ensuring that the package has the correct attributes prior to filling, e.g., color, absence of damage, absence of foreign bodies
Contents	Required by law to deliver the correct amount of beer in each container
Package integrity	Ensuring that the package is suitable for leaving the brewery, e.g., properly sealed, properly labeled (including coding)
Beer composition	See Table 7-8; most notably, content of gases

[a] Courtesy C. W. Bamforth—© ASBC.

Table 7-8.

Measurements on Beer[a]

Parameter	Preferred Frequency	Comments
Flavor clearance	Every packaging run	Simple acceptability tasting on the 1–4 scale referred to in Box 7-2
Carbon dioxide	Every packaging run	
Total in-pack oxygen	Every packaging run	
Nitrogen (if deliberately introduced to impact foam and flavor)	Every packaging run	
Alcohol	Every packaging run	May be a legal requirement
Residual extract	Every tenth run	Or if the alcohol does not "marry" with the starting gravity
Vicinal diketones[b]	Every packaging run	Ensuring prior heat treatment
Color[b]	Every packaging run	
Clarity[b]	Every packaging run	
Breakdown	Monthly	Some form of forcing test to predict clarity life
Bitterness[b]	Monthly	Whenever tasting suggests a problem
Taste and aroma[b]	Every packaging run	
pH	Every packaging run	
Total acidity[b]	As required	Only for sour beers, where it is a better gauge of perceived sourness than pH
Sulfur dioxide	Every packaging run	Only for beers that need to be labeled and/or display a flavor defect in this area
Iron[b]	As required	If metallic character is detected by taste
Foam stability[b]	Monthly	And if there are concerns
Esters[b]	Monthly	And if a concern arises during tasting
Sulfur-containing molecules[b]	Monthly	And if a concern arises during tasting
Polyphenols	As required	Only if breakdown test predicts a problem

continued

[a] Courtesy C. W. Bamforth—© ASBC.
[b] Especially on a small production scale, may simply be assessed by sight, taste, and smell.

Table 7-8. continued

Measurements on Beer[a]

Parameter	Preferred Frequency	Comments
Gushing potential[b]	Only when a problem is suspected	Should never be needed if QA approaches are in place
Gliadin	Ideally every batch	For those beers to be labeled as gluten-free or with other relevant labeling
Microbiological status	Monthly	Total cell count, unless a clear problem is indicated, in which case more comprehensive analysis is required

A Summary of the Key Points

Chapter 8

And that is all there is to it!

Quite clearly, achieving true quality in a beer is no mean feat. To conclude, here is a list of all the key take-home points that we have addressed in this book.

1. Quality is most reasonably defined as ensuring that the customers receive the beer that they thought they were going to get when they bought your beer.
2. Quality is a top-down matter. Those at the apex of the company must be committed to achieving quality and must put in place the systems to assure it.
3. People matter. All employees must be motivated, properly trained in their roles, and fully apprised of everything they need to deliver in order to fulfill the company mission.
4. All raw materials and products must be specified in sufficient detail.
5. You must have in place standard operating procedures that allow processes to be conducted consistently. These procedures must be part of a documented and comprehensive approach to quality, of the type illustrated in ISO 9001.
6. It is essential that good manufacturing practices are pursued, so too anything that is mandated by law, such as HARPC.

7. All laboratory methodology must be pursued in a consistent fashion and checked for efficacy against other laboratories. Standard methods must be used and must be validated.
8. Data should be properly plotted so that trends are clear.
9. There is no sense in measurement for measurement's sake. Perform only the analyses you need, and remember that the earlier in the process a measurement is made, the more valuable it is likely to be.
10. In relation to this, the focus should be on quality assurance and less so on quality control, although the latter is part of the former.
11. In anything other than the tiniest company, the quality assurance manager should never report to the brewmaster or anyone else responsible for making or packaging the product.
12. It is better to make measurements "on the run" in real time: tasting and smelling raw materials, monitoring the process stream and product, looking, touching, and listening (and that might be listening to people just as much as listening to pumps) as you "walk the job."
13. Keep things simple. Do not measure things that are irrelevant. Trust your suppliers on the basis that you have built up a great rapport with them. Have at least two suppliers for every ingredient.

Further Reading

The Basics of Malting and Brewing

Bamforth, C. W. (2006) Scientific Principles of Malting and Brewing. American Society of Brewing Chemists, St. Paul, MN.

Hornsey, I. (2013) Brewing. Royal Society of Chemistry, London.

Chapter 1

Bamforth, C. W. (2002) Standards of Brewing: A Practical Approach to Consistency and Excellence. Brewers Publications, Boulder, CO.

Pellettieri, M. (2015) Quality Management: Essential Planning for Breweries. Brewers Publications, Boulder, CO.

Chapter 2

Lock, D., ed. (1990) Handbook of Quality Management. Gower Publishing, Aldershot, U.K.

Chapter 3

Pripp, A. H. (2013) Statistics in Food Science and Nutrition. Springer, New York.

Chapter 4

Wheeler, D. J., and Chambers, D. S. (1992) Understanding Statistical Process Control. SPC Press, Knoxville, TN.

Chapter 5

Ishikawa, K. (1985) What is Total Quality Control? The Japanese Way. Prentice-Hall, Englewood Cliffs, NJ.

Chapter 6

Clute, M. (2009) Food Industry Quality Control Systems. CRC Press, Boca Raton, FL.

Chapter 7

Bamforth, C. W., ed. (2006) Brewing: New Technologies. Woodhead Publishing, Cambridge, U.K.

Bamforth, C. W., ed. (2016) Brewing Materials and Process: A Practical Approach to Beer Excellence. Academic Press, San Diego, CA.

Klimovitz, R., and Ockert, K., eds. (2014) Beer Packaging, 2nd ed. Master Brewers Association of the Americas, St. Paul, MN.

Quain, D. E. (2006) Yeast supply and propagation in brewing. In: Brewing: New Technologies. C. W. Bamforth, ed. Woodhead Publishing, Cambridge, U.K. pp. 167–182.

Chapter 8

Philliskirk, G. (2006) Quality. In: Handbook of Brewing, 2nd ed. F. G. Priest and G. G. Stewart, eds. CRC Press, Boca Raton, FL. pp. 729–770.

Index

Aber biomass probe, 30, 33
Perfect Pitch system, 33
accuracy, of standard methods, 22–24
acidity, of beer, 64
adjuncts, specifications and analysis sheets for, 53–54
aflatoxins, 39
agreed quality limit (AQL), 30, 32, 34
alcohol content by volume (ABV)
frequency of measurement, 64
gas chromatographic measurement, 27
taxation based on, 27
alcohol-free beers, 17–18
alkalinity, of water, 55
alpha acid, in hops, 57
American Society of Brewing Chemists (ASBC)
Laboratory Proficiency Program, 25–26
Malt-14, 26
methods of analysis, 20–21
Anheuser-Busch, 56
aroma, of beer, 37, 64
audits
internal, 12
of suppliers, 36, 49
automated feedback control, 30

bacteria. *See also* microbiological analysis; microbiological risks
dimethyl sulfide conversion in, 38
Bamforth, Charles W., 31, 35
barley
color of, 54
flaked, 54
moisture content of, 54
protein measurement in, 23–24
torrefied, 54
Bass Brewers, 3, 12, 30, 31
Preston Brook Brewery, 34, 35
beer
composition analysis of, 63, 64–65
control parameters for, 63
coolants in, 39
oxygen content, 49–50
beer tanks, pipe lines to, 34
bell curve, 18
bitterness, frequency of analysis, 64

boiling, of wort
  effect on dimethyl sulfide potential, 37
  specifications and analysis sheets for, 61
Boulton, Chris, 30, 31
breakdown, frequency of analysis of, 64
breweries, design of, 34
Brewers Association, website, 8
Brewing Industry Research Foundation, 40
Brewing Research Foundation, 40
Brewing Research Foundation International, 40
Brewing Research International, 40
bright beer chillers, coolant migration in, 40–41
bright beer tanks, oxygen levels, 44–45
Brix, 54
buildings, fitness for purpose, 38–39

calcium, water content, 55
calibration programs, 12
Campden BRI, 40
canning, package checks in, 51
carbon dioxide
  in fermentation, 62
  measurement in beer, 64
  removal, 34
  restoration to specification, 34
Central European Commission for Brewing Analysis, 20
chemical risks, identification of, 39
chloride, water content, 55
clarity
  frequency of analysis of, 64
  in lautering, 61
  prediction of, 64
  visual assessment of, 62, 63
  of wort, 62
cleaning, standard operating procedures for, 8
cleaning systems, hygiene status of, 59
codes of practice, 10
CODEX, 42
cold conditioning, 62
color
  of adjuncts, 54
  of barley, 54
  frequency of analysis of, 64
  Fusarium-related, 51
  measurement methods, 20
  of sugars, 54
Color and Clarity, 10, 58, 63
consistency, of brewing methods, 7–8, 19–20
  in boiling of wort, 61
  as brewing goal, 37, 50
  in fermentation, 30
  in systematic quality, 7–8
consortia, of brewers, 12
contractual requirements, 10–11
control diagrams and schemes, 10
coolants
  as beer contaminants, 45
  glycol as, 40, 45
  migration in bright beer chillers, 40–41
  in wort, 39
cooling, quality program for, 62
cooling towers, undesirable organisms in, 39
Cooper, Tim, 4–5
Coopers Brewery, 4–5
copper, water content, 55
corn, as adjunct, 54
costs
  preventive, 45–46
  of quality, 44–46
critical control points (CCPs), 9, 37–38. *See also* Hazard Analysis and Critical Control Points (HACCP)
  in beer transfer to bright beer chillers, 40–41
  identification of, 39
customer complaints, handling of, 12

data plotting, 68
deoxynivalenol (DON), 29–30, 51
dextrinizing units, 51
diastatic powder, 51
dimethyl sulfide (DMS), 37
  control of, 61
  critical control points, 37–38
  loss in fermenters, 38
  measurement of, 20, 26
documentation
  as HACCP program component, 39
  role in quality assurance, 9–10, 67
  uniform use of, 10
Dumas, Jean-Baptiste, 23

employees
  responsibility for quality, 3–5
  training programs for, 12, 36, 39, 46, 67
enzymes, use in gluten-reduced products, 61
equipment
  cost of, 46
  fitness for purpose, 38–39
errors, sources of, 24
esters, 62, 64
European Brewery Convention (EBC), standard methods of, 20–21

failure
  in beer gas content control, 34–36
  external and internal, 45
feeler gauges, 58, 60
fermentability, 51
fermentation
  consistency in, 20, 30
  rate, 62
  temperature for, 62
  yeast count in, 30
fermenters, 62
  dimethyl sulfide loss in, 38
filtration, 63
fitness for purpose, 38–39
Flavor, 20, 37–38, 56, 57
flavor, of beer
  acceptability tasting, 64
  dimethyl sulfide effects, 61
Foam, 30, 57
foam, stability of, 34, 64
Food and Drug Administration (FDA), 41, 42
  good manufacturing practices requirement, 8
food safety hazards
  in malt, 51
  types of, 43
food safety hazards controls
  Hazard Analysis and Critical Control Points (HACCP), 9, 10, 38–41, 42, 43
  Hazard Analysis and Risk-Based Preventive Controls (HARPC), 41–43
Food Safety Modernization Act (FSMA), 8–9, 42, 43
foreign bodies, in beer cans and bottles, 12, 63
friability, of malt, 51, 52
Fusarium, 29–30, 51

gas chromatographic measurement, 13
  of alcohol content by volume (ABV), 27
gases, beer content of. *See also* carbon dioxide; oxygen
  frequency of analysis, 63, 64
  specification targets for, 34–36
glass, safety risk from, 39
gliadin, 65
β-glucan, 51
β-glucanase, 51
gluten-reduced products, 61, 65
glycol, 40, 45
good agreement, in standard methods, 21–24
good manufacturing practices (GMP), 8–9, 42
gravity. *See also* specific gravity
  for liquid flow, 34

Index

grist, 60
  milled, 58
Grossman, Ken, 4–5
gushing/gushing potential, 29–30, 45, 65

Hazard Analysis and Critical Control Points (HACCP), 9, 38–41
  comparison with HARPC, 41, 42, 43
  definition, 38
  implementation requirements, 38–41
Hazard Analysis and Risk-Based Preventive Controls (HARPC), 41–43, 46
  comparison with HACCP, 41, 42, 43
haze, measurement of, 20–21
hefeweissen beer, 2
Hegarty, Paul, 31
hemocytometers, 59
Hershey, Milton S., 1
hops and hop extracts
  sensory testing of, 56
  specifications and analysis sheets for, 57
hydrophobin, 30
hydrostatic pressure, in fermenters, 62
hygiene
  Hazard Analysis and Critical Control Points (HACCP) guidelines, 39
  quality programs for, 58, 59

in-line analysis, 30
Institute of Brewing, 20
International Organization for Standardization (ISO)
  9000 program, 10
  9001 program, 10–12, 38
  accreditation by, 36
iodine, 60–61
ions, water content, 55
iron
  beer content, 64
  water content, 55
iso-α-acid, 57
isoamyl acetate, 25–26
iso-cohumulone, 57

Jackson, Gordon, 40, 45
Journal of the American Society of Brewing Chemists, 21

Kell, Douglas, 30, 32
Kjeldahl, Johan, 23
Kolbach index (soluble nitrogen ratio or S/T value), 51

laboratories, performance evaluation of, 25–26, 68
lagers, 37
Large, Geoff, 31
lautering
  prolonged, 45
  quality program for, 61
leadership
  commitment to quality, 3–5, 10, 67
  commitment to safety, 38
legal requirements, for quality programs, 10–11, 63, 64, 67

magnesium, water content, 55
malt
  absence of taints in, 51, 52, 53
  age, 53
  ale-type, 52
  black, 53
  black barley, 53
  caramel, 53
  color, 52, 53
  dimethyl sulfide content, 37
  free amino nitrogen content, 52
  friability, 51, 52
  homogeneity, 52
  lager-type, 52

S-methylmethionine content, 26–27, 37, 52
milled, 60
moisture content, 52, 53
oxalic acid content, 21, 24–25
pale, 52–53
representative sampling of, 50–51
roasted barley, 53
sensory testing of, 56
specialty, 53
specifications and analysis sheets for, 48–49, 51–53
undesirable components, 39
wheat, 53
manganese, water content, 55
Manners, Robin, 3, 10
mash, starch removal from, 60–61
mashing, quality program for, 60–61
Master Brewers Association of the Americas, website, 8
mean (*m*), 13–14, 17–18, 24–25
mean absolute deviation (MAD), 15
mean square deviation (MSD), 16
measurements, for quality assurance, 29–36. *See also* standard methods; statistics
accuracy of, 22–24
of beer components, 64–65
categories of, 29–30
"in specification," 32
precision of, 26–27
real-time, 30, 64
shifts in, 27
timing of, 29–30, 48, 68
types of, 29, 48
meetings, for quality review, 11–12
methods of analysis, 20–21
methyl blue stain, for yeast, 59
microbiological analysis
frequency of performance, 65
of raw materials, 59
microbiological risks, identification of, 39
mill settings, 58, 60
Mittteleuropäische Brautechnische Analysenkommission (MEBAK), 20
moisture content
of barley, 54
of malt, 52, 53

National Advisory Committee on Microbiological Criteria for Foods, 42
National Primary Drinking Water Regulations, 49, 54–55
nitrogen, measurement in beer, 64
normal distribution, 17–18, 24–25

oxalic acid, measurement in malt, 21, 24–25
oxygen
addition during cooling, 62
beer content, 49–50
levels during fermentation, 62
levels in bright beer tanks, 44–45
purging of, 34, 44–45
total in-pack, 64
oxygen analyzers, 62

packaged beer, key control parameters, 63
packaging, 63
pH
in fermentation, 62
frequency of analysis of, 64
in lautering, 61
as sourness indicator, 64
pipe lines, 34
polyphenols, 64
polyvinylpyrrlidone (PVPP), 10
precision
of measurement, 26–27
of standard methods, 22–24
pressure differentiation, during beer transfer, 40–41
preventive costs, 45–46
proactive techniques, 36, 48

processes, standard operating procedures for, 8, 9
process flow charts, 39
product-recall systems, 39
prolyl endoproteinase (PEP), 61
protein, barley content, 23–24
purchase, of materials and services, 12. *See also* suppliers

Quain, David, 30, 31
quality. *See also* systematic quality
  cost of, 44–45
  definition, 1–3, 67
  organizational factors in, 3–5
  production expediency versus, 49–50
  top-down approach to, 3–5, 10, 67
quality assurance, 29–36
  comparison with quality control, 34–36, 68
  cost of, 45–46
  mission of, 35–36
  as proactive process, 36, 48
quality assurance managers, 47
  reporting hierarchy of, 49–50
quality control, comparison with quality assurance, 34–36
quality manuals, 7–8, 10
quality policy, 10, 11
quality programs, components of, 47–65. *See also* specifications and analysis sheets
  adjuncts, 53–54
  baseline approach, 50–65
  beer measurements, 63, 64–65
  boiling of wort, 61
  cold conditioning, 62
  cooling, 62
  fermentation, 62
  filtration, 63
  hops, 57
  hygiene, 58, 59
  lautering, 61
  legal aspects, 10–11, 63, 64, 67
  malt, 51–53
  market-protocol match, 47–48
  mashing, 60–61
  mill settings, 58, 60
  packaging, 63
  principles of, 47–50
  stabilization, 63
  warehousing and storage, 63
  water, 54–56
  whirlpools, 61
  yeast, 58, 59

range (statistical), 14–16
raw materials
  analysis of, 29–30, 48–49
  microbiological evaluation of, 59
  representative sampling of, 50–51
  specifications for, 67
  transport of, 39
real-time measurements, 30
reject quality limit (RQL), 30, 32, 34
reliability, 25
repeatability (*r*), 26–27
representative sampling
  of malt, 50–51
  of pitching wort, 62
reproducibility (*R*), 26–27
residual extract, 64
reviews, of quality systems, 11–12
rice, as adjunct, 54
ring analysis, 36
roller mills, 9
  settings of, 58, 60

safety products, commitment to, 38
safety. *See also* food safety hazards controls
  identification of risks to, 38–39
sensory testing
  of beer, 64
  of hops and hop teas, 56
  of malt, 56
  of water, 55–56
  of wort, 56
Sierra Nevada (beer company), 4–5
silicate, water content, 55

six sigma range, 32
S-methylmethionine (SMM)
conversion to dimethyl sulfide, 20, 37, 38, 61
degradation, 37
in malt, 26–27
precision of measurement of, 26
Snake Venom beer, 17
sodium, water content, 55
soluble nitrogen ratio (Kolbach index or S/T value), 51
sour beer, 64
specifications
beer oxygen content, 49–50
ISO requirements, 10–12
relationship to standard methods, 24
"right the first time," 35–36
specifications and analysis sheets
adjuncts, 53–54
hops and hop extracts, 57
malt, 48–49, 51–53
water quality, 54–56
yeast, 58, 59
specific gravity
in boiling of wort, 61
in fermentation, 62
in lautering, 61
stabilization, efficacy evaluation of, 63
staff. *See* employees
standard deviation (σ), 16–17, 24–25
95%, 18, 24–25, 32
99%, 18, 32
of agreed quality limit (AQL), 32
of reject quality limit (RQL), 32
standard methods, 19–27
accuracy and precision of, 22–24
establishment of, 21
good agreement in, 21–24
groups of, 20–21
relationship to specifications, 24
repeatability (*r*), 26–27
reproducibility (*R*), 26–27
Youden plots, 24–26, 32
standard operating procedures (SOP), 8, 9
adjustment for problem recurrence prevention, 12
maintenance of, 36
for mashing, 60
starch, removal from mash, 60–61
statistics, 13–18
bell curves, 18
mean (*m*), 13–14
mean absolute deviation (MAD), 15
mean square deviation (MSD), 16
normal distribution, 17–18, 24–25
range, 14–16
reliability of, 25
standard deviation (σ), 16–17, 18, 24–25, 32
steep tanks, 49
storage, 63
S/T value (Kolbach index or soluble nitrogen ratio), 51
sugars, 54
sulfate, water content, 55
sulfur-containing molecules, 64
sulfur dioxide, frequency of analysis of, 64
suppliers, 12
audits of, 36, 49
relationship with, 49, 51, 68
safety requirements for, 39
of water, 54–55
syrups, 54
systematic quality, 7–12
consistency in, 7–8
Food Safety Modernization Act (FSMA), 8–9, 42, 43
good manufacturing practices (GMP), 8–9
Hazard Analysis and Critical Control Points (HACCP), 9
Hazard Analysis and Risk-Based Preventive Controls (HARPC), 9
ISO 9000, 10

ISO 9001, 10–12, 38
reviews of quality systems, 11–12
standard operating procedures (SOP), 8, 9, 12, 36, 60

taste testing. *See* sensory testing
taxation, based on alcohol content by volume (ABV), 27
temperature
in cold conditioning, 62
in fermentation, 62
in warehousing and storage, 63
of water, in mashing, 60
terrorism, 43
Todd, Bob, 30
total dissolved solids, water content, 55
total protein, 51
total quality management (TQM), 36
total soluble protein (nitrogen) (TSN), 48
training, of employees, 12, 36, 39, 67
cost of, 46
turbidity
quantification of, 63
of water, 55

U.S. Department of Agriculture, 41, 42

viability, of yeast, 59, 62
vicinal diketones, 56, 62, 64
viscosity, of wort, 51
volatile compounds, suppression of, 62

warehousing, 63
water
analysis of, 30, 49
quality monitoring, 54–56
sensory testing of, 55–56
suppliers of, 54–55
water temperature, in mashing, 60
wheat, as adjunct, 54
whirlpools
effect on dimethyl sulfide content, 37
quality program for, 61
White, Harry, 34
wort
clarity, 62
coolants in, 39
properties of, 60
sensory testing of, 56
spoilage bacteria in, 38
strength, 62
total soluble protein (nitrogen) (TSN), 48
viscosity, 51

yeast
age, 59
premature flocculation factor, 51
viability, 59, 62
wild, 59
yeast count
with Aber biomass probe, 30
viable, 62
Youden plots, 24–26, 32
Youden, William, 24

Notes

Notes